Also by Jon Ronson

THEM: ADVENTURES WITH EXTREMISTS

THE MEN WHO STARE AT GOATS

Jon Ronson

SIMON & SCHUSTER PAPERBACKS

NEW YORK · LONDON · TORONTO · SYDNEY

SIMON & SCHUSTER PAPERBACKS
A Division of Simon & Schuster, Inc.
1230 Avenue of the Americas
New York, NY 10020

Copyright © 2004 by Jon Ronson

Originally published in Great Britain in 2004 by Picador,
an imprint of Pan Macmillan Ltd.

All rights reserved, including the right to reproduce this book or
portions thereof in any form whatsoever. For information, address
Simon & Schuster Paperbacks Subsidiary Rights Department,
1230 Avenue of the Americas, New York, NY 10020.

This Simon & Schuster trade paperback edition October 2009

SIMON & SCHUSTER PAPERBACKS and colophon are registered
trademarks of Simon & Schuster, Inc.

For information about special discounts for bulk purchases,
please contact Simon & Schuster Special Sales at
1-866-506-1949 or business@simonandschuster.com.

The Simon & Schuster Speakers Bureau can bring authors
to your live event. For more information or to book an event,
contact the Simon & Schuster Speakers Bureau at
1-866-248-3049 or visit our website at www.simonspeakers.com.

Manufactured in the United States of America

10 9 8 7 6 5 4

Library of Congress Control Number: 2005041263

ISBN 978-0-7432-4192-2
ISBN 978-1-4391-8177-5 (pbk)

For John Sergeant and also for General Stubblebine

CONTENTS

CONTENTS

The Men Who Stare at Goats

1. THE GENERAL

This is a true story. It is the summer of 1983. Major General Albert Stubblebine III is sitting behind his desk in Arlington, Virginia, and he is staring at his wall, upon which hang his numerous military awards. They detail a long and distinguished career. He is the United States Army's chief of intelligence, with sixteen thousand soldiers under his command. He controls the army's signals intelligence, their photographic and technical intelligence, their numerous covert counterintelligence units, and their secret military spying units, which are scattered throughout the world. He would be in charge of the prisoner-of-war interrogations too, except this is 1983, and the war is cold, not hot.

He looks past his awards to the wall itself. There is something he feels he needs to do even though the thought of it frightens him. He thinks about the choice he has to make. He can stay in his office or he can go into the next office. That is his choice. And he has made it.

He is going into the next office.

General Stubblebine looks a lot like Lee Marvin. In fact, it is widely rumored throughout military intelligence that he is

Lee Marvin's identical twin. His face is craggy and unusually still, like an aerial photograph of some mountainous terrain taken from one of his spy planes. His eyes, forever darting around and full of kindness, seem to do the work for his whole face.

In fact he is not related to Lee Marvin at all. He likes the rumor because mystique can be beneficial to a career in intelligence. His job is to assess the intelligence gathered by his soldiers and pass his evaluations on to the deputy director of the CIA and the chief of staff for the army, who in turn pass it up to the White House. He commands soldiers in Panama, Japan, Hawaii, and across Europe. His responsibilities being what they are, he knows he ought to have his own man at his side in case anything goes wrong during his journey into the next office.

Even so, he doesn't call for his assistant, Command Sergeant George Howell. This is something he feels he must do alone.

Am I ready? he thinks. Yes, I am ready.

He stands up, moves out from behind his desk, and begins to walk.

I mean, he thinks, what is the atom mostly made up of anyway? Space!

He quickens his pace.

What am I mostly made up of? he thinks. Atoms!

He is almost at a jog now.

What is the wall mostly made up of? he thinks. Atoms! All I have to do is merge the spaces. The wall is an illusion. What is destiny? Am I destined to stay in this room? Ha, no!

Then General Stubblebine bangs his nose hard on the wall of his office.

Damn, he thinks.

General Stubblebine is confounded by his continual failure to walk through his wall. What's wrong with him that he can't do it? Maybe there is simply too much in his in-tray for him to give it the requisite level of concentration. There is no doubt in his mind that the ability to pass through objects will one day be a common tool in the intelligence-gathering arsenal. And when that happens, well, is it too naive to believe it would herald the dawning of a world without war? Who would want to screw around with an army that could do *that*? General Stubblebine, like many of his contemporaries, is still extremely bruised by his memories of Vietnam.

These powers *are* attainable, so the only question is, by whom? Who in the military is already geared toward this kind of thing? Which section of the army is trained to operate at the peak of their physical and mental capabilities?

And then the answer comes to him.

Special Forces!

This is why, in the late summer of 1983, General Stubblebine flies down to Fort Bragg, in North Carolina.

Fort Bragg is vast—a town guarded by armed soldiers, with a mall, a cinema, restaurants, golf courses, hotels, swimming pools, riding stables, and accommodations for forty-five thousand soldiers and their families. The general drives past these places on his way to the Special Forces Command Center. This is not the kind of thing you take into the mess hall. This is for Special Forces and nobody else. Still, he's afraid. What is he about to unleash?

In the Special Forces Command Center, the general decides to start soft. "I'm coming down here with an idea," he begins.

The Special Forces commanders nod.

"If you have a unit operating outside the protection of mainline units, what happens if somebody gets hurt?" he says. "What happens if somebody gets wounded? How do you deal with that?"

He surveys the blank faces around the room.

"Psychic healing!" he says.

There is a silence.

"This is what we're talking about," says the general, pointing to his head. "If you use your mind to heal, you can probably come out with your whole team alive and intact. You won't have to leave anyone behind." He pauses, then adds, "Protect the unit structure by hands-off and hands-on healing!"

The Special Forces commanders don't look particularly interested in psychic healing.

"Okay," says General Stubblebine. The reception he's getting is really quite chilly. "Wouldn't it be a neat idea if you could teach somebody to do *this*?"

General Stubblebine rifles through his bag and produces, with a flourish, bent cutlery.

"What if you could do this?" says General Stubblebine. "Would you be interested?"

There is a silence.

General Stubblebine finds himself beginning to stammer a little. They're looking at me as if I'm nuts, he thinks. I am not presenting this correctly.

He glances anxiously at the clock.

"Let's talk about time!" he says. "What would happen if time is not an instant? What if time has an X-axis, a Y-axis, and a Z-axis? What if time is not a point but a space? At any particular time we can be *anywhere* in that space! Is the space confined to the ceiling of this room, or is the space *twenty million miles?*" The general laughs. "Physicists go *nuts* when I say this!"

Silence. He tries again.

"Animals!" says General Stubblebine.

The Special Forces commanders glance at one another.

"Stopping the hearts of animals," he continues. "*Bursting* the hearts of animals. This is the idea I'm coming in with. You have access to animals, right?"

"Uh," say Special Forces. "Not really . . ."

General Stubblebine's trip to Fort Bragg was a disaster. It still makes him blush to recall it. He ended up taking early retirement in 1984. Now, the official history of army intelligence, as outlined in their press pack, basically skips the Stubblebine years, 1981–84, almost as if they didn't exist.

In fact, everything you have read so far has for the past two decades been a military intelligence secret. General Stubblebine's doomed attempt to walk through his wall and his seemingly futile journey to Fort Bragg remained undisclosed right up until the moment that he told me about them in room 403 of the Tarrytown Hilton, just north of New York City, on a cold winter's day two years into the War on Terror.

"To tell you the truth, Jon," he said, "I've pretty much blocked the rest of the conversation I had with Special Forces out of my head. Whoa, yeah. I've *scrubbed* it from my mind! I walked away. I left with my tail between my legs."

He paused, and looked at the wall.

"You know," he said, "I really thought they were great ideas. I still do. I just haven't figured out how *my* space can fit through *that* space. I simply kept bumping my nose. I couldn't . . . No. *Couldn't* is the wrong word. I never got myself to the right state of mind." He sighed. "If you really want to know, it's a disappointment. Same with the levitation."

Some nights, in Arlington, Virginia, after the general's first wife, Geraldine, had gone to bed, he would lie down on his living-room carpet and try to levitate.

"And I failed totally. I could not get my fat ass off the ground, excuse my language. But I still think they were great ideas. And do you know why?"

"Why?" I asked.

"Because you *cannot* afford to get stale in the intelligence world," he said. "You *cannot* afford to miss something. You don't believe that? Take a look at terrorists who went to flying schools to learn how to take off but not how to land. And where did *that* information get lost? You *cannot* afford to miss something when you're talking about the intelligence world."

There was something about the general's trip to Fort Bragg that neither of us knew the day we met. It was a piece of information that would soon lead me into what must be among the most whacked-out corners of George W. Bush's War on Terror.

What the general didn't know—what Special Forces kept secret from him—was that they actually considered his ideas to be excellent ones. Furthermore, as he proposed his clandestine animal-heart-bursting program and they told him that they didn't have access to animals, they were concealing the fact that there were a hundred goats in a shed just a few yards down the road.

The existence of these hundred goats was known only to a select few Special Forces insiders. The covert nature of the goats was helped by the fact that they had been de-bleated; they were just standing there, their mouths opening and closing, with no bleat coming out. Many of them also had their legs bandaged in plaster.

This is the story of those goats.

2. GOAT LAB

It was Uri Geller who set me on the trail that led to the goats. I met him on the roof terrace of a central London restaurant in early October 2001, less than a month into the War on Terror. There had long been rumors (circulated on the whole, it must be said, by Uri himself) that back in the early 1970s he had been a psychic spy working secretly for U.S. intelligence. Many people have doubted his story—*The Sunday Times* once called it "a bizarre claim," arguing that Uri Geller is nuts whereas the intelligence establishment is not. The way I saw it, the truth lay in one of four possible scenarios:

1. It just never happened.

2. A couple of crazy renegades in the higher levels of the U.S. intelligence community had brought in Uri Geller.

3. U.S. intelligence is the repository of incredible secrets, which are kept from us for our own good; one of those secrets is that Uri Geller has psychic powers, which were

harnessed during the Cold War. They just hoped he wouldn't go around telling everybody.

4. The U.S. intelligence community was, back then, essentially nuts through and through.

Uri was quiet in the restaurant. He wore big, wraparound mirrored sunglasses. His brother-in-law, Shipi, was equally unforthcoming, and the whole thing was a bit awkward. I had met them once or twice before and had found them to be infectiously ebullient people. But there was no ebullience that day.

"So," I said, "let's start. How did you first become a psychic spy for the U.S. government?"

There was a long silence.

"I don't want to talk about it," Uri said.

He sipped his mineral water and glanced over at Shipi.

"Uri?" I said. "What's wrong? You *often* talk about it."

"No I don't," he said.

"Yes you *do*!" I said.

I had been researching this for two weeks, and I had already amassed a file an inch thick of his reminiscences about his psychic spying days, dictated to journalists throughout the 1980s and 1990s, who then added sarcastic asides. In more or less every article, the line of reasoning was the same: The intelligence services wouldn't *do* that. There was an almost frantic reluctance to accept Uri's word, or even to make a few calls to verify or refute it. For all our cynicism, we apparently still invested the intelligence services with some qualities of rigor and scientific methodology. The few journalists who accepted Uri's claim implicitly expressed relief that all this happened a long time ago, back in the 1970s.

"I never talk about it," Uri said.

"You spoke about it to the *Financial Times,*" I said. "You said you did a lot of psychic work for the CIA in Mexico."

Uri shrugged.

A plane flew low overhead and everyone on the terrace stopped eating for a moment and looked up. Ever since 9/11, Attorney General John Ashcroft had been warning of imminent terrorist attacks—on banks, apartment blocks, hotels, restaurants, and shops in the United States. On one occasion President Bush announced that he couldn't say *anything at all* about a particular looming cataclysm. Equally nonspecific high alerts were occurring in London too. Then, suddenly, Uri took off his sunglasses and looked me squarely in the eye.

"If you repeat what I am about to tell you," he said, "I will deny it."

"Okay," I said.

"It will be your word against mine," said Uri.

"Okay," I said.

Uri moved his chair closer to mine. He glanced around the restaurant.

"This," he said, "is no longer a history story."

"I'm sorry?" I said.

"I have been reactivated," said Uri.

"What?" I said.

I looked over at Shipi. He nodded gravely.

"I don't suppose it was you who told John Ashcroft about the hotels and the banks and the apartment blocks?" I asked.

"I am saying nothing else," said Uri.

"Uri," I said, "please give me something to go on. Please tell me one more thing."

Uri sighed.

"Okay," he said. "I will tell you one more thing only. The man who reactivated me is . . ." Uri paused, then he said, "called Ron."

And that was it. I have not spoken to Uri Geller since. He has not returned my calls. He refused to divulge anything further about Ron. Was Ron FBI? CIA? military intelligence? Homeland Security? Could Ron be MI5? MI6? Was Uri Geller playing a part in the War on Terror?

I had a minor breakthrough a year later, in a hotel in Las Vegas, when I was interviewing one of General Stubblebine's former military spies, Sergeant Lyn Buchanan. I said, "Uri Geller says that the man who reactivated him is called Ron." Sergeant Buchanan fell silent and then he nodded enigmatically and said, "Ah, Ron. Yes. I know Ron."

But he wouldn't tell me anything more about him.

General Stubblebine wouldn't talk about Ron either.

"The damn psychic spies should be keeping their damn mouths shut," he said, "instead of chitchatting all over town about what they did."

The general, I discovered in the weeks after I met Uri, had commanded a secret military psychic spying unit between 1981 and 1984. The unit wasn't quite as glamorous as it might sound, he said. It was basically half-a-dozen soldiers sitting inside a heavily guarded, condemned clapboard building in Fort Meade, Maryland, trying to be psychic. Officially the unit did not exist. The psychics were what is known in military jargon as Black Ops. Because they didn't "exist" they were not permitted access to the army's coffee budget. They had to bring their own coffee into work. They had come to resent this. Some of them were in there, trying to be psychic, from 1978 until 1995. From time to time, one of them died or went stir-crazy, and a new psychic soldier would be brought in to replace the casualty. When one of them got a vision of a Russian warship, or a future event he would sketch it, and pass the sketches up the chain of command.

And then, in 1995, the CIA closed them down.

Many of the psychic soldiers have subsequently published their autobiographies, such as *The Seventh Sense: The Secrets of Remote Viewing as Told by a "Psychic Spy" for the U.S. Military,* by Lyn Buchanan.

"Everybody wants to be the first on the publicity stump," said General Stubblebine. "I could wring some of their necks."

And that was all the general would say about the psychic spies.

"Are they back in business?" I asked him.

"I hope so," he said.

"Was Uri Geller one of yours?" I asked.

"No," he said, "but I wish he had been. I am a great fan of his."

And so it was that my quest to track down Ron took me to Hawaii, to a house on the road between Honolulu and Pearl Harbor, the home of retired Sergeant First Class—and onetime Special Forces psychic spy—Glenn Wheaton. Glenn was a big man with a tight crop of red hair and a Vietnam-vet-style handlebar mustache. My plan was to ask Glenn about his psychic spying days and then try to broach the subject of Ron, but from the moment I sat down, our conversation veered off in a wholly unexpected direction.

Glenn leaned forward in his chair. "You've gone from the front door to the back door. How many chairs are in my house?"

There was a silence.

"You probably can't tell me how many chairs are in my house," said Glenn.

I started to look around.

"A supersoldier wouldn't need to look," he said. "He would just *know*."

"A supersoldier?" I asked.

"A supersoldier," said Glenn. "A Jedi Warrior. He would know where all the lights are. He would know where all the power outlets are. Most people are poor observers. They haven't got a clue about what's really happening around them."

"What's a Jedi Warrior?" I asked.

"You're looking at one," said Glenn.

In the mid-1980s, he told me, Special Forces undertook a secret initiative, codenamed Project Jedi, to create supersoldiers—soldiers with superpowers. One such power was the ability to walk into a room and instantly be aware of every detail; that was level one.

"What was the level above that?" I asked.

"Level two," he said. "Intuition. Is there some way we can develop you so you make correct decisions? Somebody runs up to you and says, 'There's a fork in the road. Do we turn left or do we turn right?' And you go"—Glenn snapped his fingers—" 'We go right!' "

"What was the level above that?" I asked.

"Invisibility," said Glenn.

"Actual invisibility?" I asked.

"At first," said Glenn. "But after a while we adapted it to just finding a way of *not being seen*."

"In what way?" I asked.

"By understanding the linkage between observation and reality, you learn to dance with invisibility," said Glenn. "If you're not observed, you are invisible. You only exist if someone sees you."

"So, like camouflage?" I asked.

"No," sighed Glenn.

"How good are you at invisibility?" I asked.

"Well," said Glenn, "I've got red hair and blue eyes, so people tend to remember me. But I get by. I'm alive today."

"What was the level above invisibility?" I asked.

"Uh," said Glenn. He paused for a moment. Then he said, "We had a master sergeant who could stop the heart of a goat."

There was a silence. Glenn raised an eyebrow.

"Just by . . ." I said.

"Just by *wanting* the goat's heart to stop," said Glenn.

"That's quite a leap," I said.

"Right," said Glenn.

"And did he make the goat's heart stop?" I asked.

"He did it at least once," said Glenn.

"Huh," I said. I really didn't know how to respond to this. "But it's not really an area you want to . . ."

"Go to," I said.

"That's right," said Glenn. "Not an area you want to go to, because as it turned out in the evaluation he actually did some damage to himself as well."

"Huh," I said again.

"Sympathetic injury," said Glenn.

"So it's not as if the goat was psychically fighting back?" I asked.

"Goat didn't have a chance," said Glenn.

"Where did this happen?" I asked.

"Down in Fort Bragg," he said, "at a place called Goat Lab."

"Glenn," I said, "will you tell me everything about Goat Lab?"

And so Glenn began.

Goat Lab, which exists to this day, is secret. Most of the soldiers who live and work within Fort Bragg don't even know of its existence. Those military personnel not in the loop, said Glenn, assume that the rickety clapboard hospital buildings dating from the Second World War, situated down an unpaved track in an overgrown wooded area, are derelict. In fact, they are filled with one hundred de-bleated goats.

The goats weren't covertly herded into these buildings just so the Jedi Warriors could stare at them. Goat Lab was originally created as a clandestine laboratory to provide in-the-field surgical training for Special Forces soldiers. During this more conventional phase of the goats' lives, each one was taken through a heavy steel soundproofed door into a bunker and shot in the leg using a bolt gun. Then the Special Forces trainees would rush the goat into an operating theater, anesthetize it, dress the wound, and nurse it back to health. Goat Lab used to be called Dog Lab, but it turned out that nobody wanted to do all that to dogs, so they switched to goats. It was apparently determined within Special Forces that it was just about impossible to form an emotional bond with a goat. In fact, according to People for the Ethical Treatment of Animals (PETA), goats have historically made up an unusually large percentage of the estimated million animals on the receiving end of covert experiments within the army. Most goat-related military activity remains highly classified, but from time to time some details have leaked out. When an atomic bomb was detonated in the sky near Bikini Atoll in the South Pacific in 1946, for example, most of the four thousand animals that had been dispatched by the military to float around underneath the explosion on a boat known as the Atomic Ark were goats. They wanted to see how the animals would fare with the fallout. They fared badly. Additionally, several thousand goats are currently being transformed—on an air force base—into a weird kind of goat/spider hybrid. "Spider silk is truly a prized biomaterial that's really been kept from mankind simply because, up until now, only spiders can make it," an air force spokesperson explained to CBC news in Canada. "Once a spider-cell gene has actually become part of the goat's genetic makeup, that goat will produce spider silk on a very cost-effective basis for many years to come. The magic is in their milk. A single gram of it will produce thousands of meters of silk thread that can be woven into bulletproof vests for tomorrow's military."

And now there was the work undertaken inside Goat

Lab—the de-bleatings and shootings and so on. Could all this, I wondered, explain how a master sergeant had managed to kill a goat just by staring at it? Perhaps, before he got to his goat, it was already in shaky medical condition: some goats were recovering amputees; others had been cut open, had their hearts and kidneys scrutinized, and were then closed up again. Even the luckier goats—the ones that had only been shot—were presumably hobbling around Goat Lab in eerie silence with their legs in plaster. Perhaps the master sergeant had been staring at a particularly sickly goat? But Glenn Wheaton said he couldn't remember anything about the health of the goat in question.

"How did the master sergeant get sick as a result of stopping its heart?" I asked.

"To generate enough power," Glenn replied, "to generate enough force of intent to damage the goat, he damaged himself. Everything goes with a cost, see? You pay the piper."

"What part of him got damaged?" I asked.

"*His* heart."

"Huh," I said.

There was a silence.

"Can *you* stop a goat's heart?" I asked Glenn.

"*No!*" said Glenn, startled. "No! No, no, no!"

Glenn looked around him, as if he was afraid that the very question might implicate him in the act and put him in the bad books of some unseen spiritual force.

"Do you just not *want* to do it?" I asked. "*Do* you have the power to stop a goat's heart?"

"No," said Glenn. "I don't think I have the power to stop a goat's heart. I think that if one trained oneself to get to that level, one would have to say, 'What did the goat ever do to me? *Why that goat?*'"

"So who did achieve that level?" I asked. "Who was the master sergeant?"

"His name," said Glenn, "was Michael Echanis."

And that, said Glenn, was all he knew about Goat Lab.

"Glenn," I said, "are goats being stared at once again post-September 11?"

Glenn sighed.

"I'm out of the military," he replied. "I'm out of the loop. I know no more than you do. If I phoned Special Forces I'd get the same response they'd give you."

"Which is what?"

"They would neither confirm nor deny. The very existence of the goats is hush-hush. They won't even admit to *having* goats."

This, I later learned, was the reason for the de-bleating. It was done not because the Special Forces soldiers were required to learn how to cauterize the vocal cords of the enemy, but because Special Forces were concerned that a hundred bleating goats on base could come to the attention of the local ASPCA.

Glenn was looking a little panicked. "This is Black Op stuff," he said.

"Where can I go from here?" I asked.

"Nowhere," said Glenn. "Forget it."

"I can't forget it," I said. "It is an image I am unable to get out of my head."

"Forget it!" said Glenn. "Forget I ever said anything about the goats."

But I couldn't. I had many questions. For instance, how did all this begin? Did Special Forces simply steal General Stubblebine's idea? It seemed possible, given the timeline I was beginning to piece together. Perhaps Special Forces feigned chilly indifference to the general's animal-heart-bursting initiative and subsequently instructed Michael Echanis, whoever he was, to begin staring. Maybe they simply wanted the glory for themselves in the event that staring an enemy to death became a tool in the military arsenal and changed the world forever.

Or was it a *coincidence*? Were Special Forces, unknown to General Stubblebine, working on the goats already? The answer to this question, I felt, might provide some insight into the U.S. military mind. Is this the kind of idea that people *routinely* have in those circles?

After I left Glenn Wheaton I tried to find out everything I could about Michael Echanis. He was born in 1950, in Nampa, Idaho. The old lady who lived down the street from him was "a real grouch," according to a childhood friend, "so Michael blew up her woodshed."

He fought in Vietnam for two months in 1970, during which time he shot twenty-nine people—"confirmed kills"—but then parts of his foot and calf were blown off and he was shipped back home to San Francisco, where the doctors told him he'd never walk again. But he confounded them and by 1975 had instead become the nation's leading exponent of the Korean martial art of Hwa Rang Do, teaching such techniques as invisibility to Special Forces at Fort Bragg.

"If you have to be by a wall with horizontal brickwork, don't stand vertically," he'd tell his Green Beret trainees. "In a tree, try to look like a tree. In open spaces, fold up like a rock. Between buildings, look like a connecting pipe. If you need to pass along a featureless white wall, use a reversible piece of cloth. Hold up a white square in front of you as you move. Think black. That is the nothingness."

This nothingness was important to Echanis. Within that nothingness, he found that he could kill. A former martial arts colleague of Echanis's named Bob Duggan once told *Black Belt* magazine that he considered Echanis basically psychotic. He said Echanis was always on the verge of creating mayhem, always thinking about death and the process of death, and that this character trait had lodged itself in Echanis's psyche around the time of his twenty-nine confirmed Vietnam kills and the subsequent blowing off of his foot.

"Look at your target's arms or legs," Echanis would tell his Green Beret trainees. "Don't look at his eyes until the last second. You can freeze a person by locking on to him with your eyes for a split second. I walk up to a person not looking at him, suddenly I look intently at him. As our eyes make contact, he looks at me. At that split second, his body is frozen, and that is when I hit him. You can talk smoothly. Go into a monotone. 'No, I'm not going to stab or attack you.' Then do it. If you're totally relaxed in eyes, body, voice, it will not occur to the other person that you are ready to move on him."

In the mid-1970s, Echanis published a book titled *Knife Self-Defense for Combat,* which advocated the controversial technique of noisily leaping in the air and spinning around while attacking an enemy with a knife. This approach was hailed by some knife-fighting aficionados but criticized by others who believed that the leaping and the spinning might lead one accidentally to stab oneself, and that one should keep one's footwork simple when armed with a knife.

Nonetheless, Echanis's superpowers became the stuff of legend. One former Green Beret reminisced on the Internet:

> I was open-mouthed and slack-jawed. I watched as he lay on a bed of nails while a trainee broke a cinder block on his stomach with a sledgehammer, he put steel spokes through the skin of his neck and forearms and lifted buckets of sand, then removed them with no bleeding and very little physical evidence of trauma, he had a tug-of-war with a dozen men who could not budge him a single inch, he even hypnotized a couple of the people in attendance. Green Berets were tossed around like rag dolls. The pain he could inflict was surreal. He could hurt someone badly with a finger. Mike, you're not forgotten. The knife you gave me lies next to my beret. You tempered my soul for life. God bless Mike Echanis!

Echanis spent some time as the martial arts editor of *Soldier of Fortune* magazine, the "journal of professional adventurers." He became something of a poster boy for mercenaries, quite literally, in fact, because he frequently appeared on the covers of *Soldier of Fortune* and *Black Belt*. If you ever chance upon a 1970s photograph of a handsome and wiry American mercenary with a handlebar mustache, lying vigilant and armed in jungle terrain, wearing khaki and a bandanna and clutching a knife with a vicious serrated edge, the chances are that it is Michael Echanis. All this made him even more famous, which is not a good strategy for a mercenary, and possibly led to his mysterious death at the age of twenty-eight.

There are a number of versions out there of how Echanis came to die. What is definite is that it happened in Nicaragua, where he had hooked up, in a professional capacity, with its then-dictator, Anastasio Somoza. Some reports say that it was the CIA who brokered the meeting between the two men, and that the agency gave Echanis a $5 million budget to teach eso-teric martial-arts techniques to Somoza's Presidential Guard and anti-Sandinista commandos.

Echanis told one Somoza biographer that the reason why he loved being in Nicaragua was that at home in the United States it was really hard to walk down the street and get into a fight. But in Nicaragua, he said, he could get into fights almost every day.

It could be argued that being paid by Somoza to help crush peasant insurrection was somewhat unheroic, but fans of Echanis told me, when I tentatively put this to them, that it made Echanis's courage all the more outstanding, as the American people were not exactly enamored of Somoza, and the press "made the Sandinistas into saints."

One version of the events surrounding Echanis's death goes like this: Echanis and a few fellow mercenaries were in a helicopter, on their way to perpetrate some Somoza-inspired horror. The helicopter exploded, either as a result of a bomb planted by anti-Somoza rebels or because the pas-sengers were messing around with grenades and one of them went off, and everyone on board was killed.

In the other version, told to me by martial arts master Pete Brusso, who teaches at the Camp Pendleton Marine Training Base in San Diego, Echanis was not in a helicopter. He was on the ground, acting too big for his boots regarding his superhuman powers.

"He used to let jeeps run over him," explained Pete Brusso. "Special Forces would get a jeep, and he'd lie down on the ground, and the jeep would go slowly over him. That isn't too hard to do. A two-thousand-five-hundred-pound jeep, four wheels, you divide that by four. If it goes slow enough over you, the body can pretty much take it. But if you hit the body with any speed, you've got a kinetic-energy shock to it."

Pete said that Echanis had challenged some fellow mercenaries to drive over him so he could prove that his fearsome reputation was warranted.

"Well, whoever was driving the jeep didn't realize he was supposed to slow down," said Pete Brusso. "Oops!" He laughed. "Yeah, so he took internal injuries and died. That's what I heard."

"Do you think they subsequently made up the helicopter story to spare everyone's embarrassment and possible legal recriminations?" I asked.

"Could be," said Pete Brusso.

But in none of the stories I read or heard about Michael Echanis could I find any reference to him killing a goat just by staring at it, so I was at a dead end regarding Goat Lab.

Strangely, in fact, whenever I broached the subject of goat staring in my e-mail exchanges with former friends and associates of Echanis, they immediately, on every single occasion, stopped e-mailing me back. I started to think that perhaps they thought I was nuts. This is why, after a while, I began avoiding crazy-sounding words like *goat* and *staring* and *death*, and instead asked questions like, "Do you happen to know whether or not Michael was ever involved in attempting to influence livestock from afar?"

But even then the e-mail exchanges abruptly halted. Perhaps I had indeed stumbled on a secret so sensitive that nobody wanted to admit to any knowledge of it.

So I called Glenn Wheaton again.

"Just tell me whose original idea the goat staring was," I said. "Just tell me that."

Glenn sighed. He said a name.

Over the next few months, other former Jedi Warriors gave me the same name. It kept coming up. It is a name few people outside the military have ever heard. But it was this man who inspired the Jedi Warriors to do what they did. In fact, this one man, armed with a passion for the occult and a belief in superhuman powers, has had a profound and hitherto unchronicled impact on almost every aspect of army life. General Stubblebine's doomed attempt to pass through his wall was inspired by this man, as was—at the other end of the scale of secrecy—the army's famous TV recruitment slogan, "Be All You Can Be."

> You're reaching deep inside of you
> For things you've never known.
> Be all that you can be
> You can do it
> In the Army.

THE MEN WHO STARE AT GOATS

This slogan was once judged by *Advertising Age* magazine to be the second most-effective jingle in the history of U.S. television commercials (the winner being "You Deserve a Break Today, So Get Up and Get Away, to McDonald's"). It touched the Reaganite soul of college graduates across the nation in the 1980s. Who would have believed that the soldier who helped inspire the jingle had such a fabulous idea of what "All You Can Be" might include?

Although this man was filled with the kindest of intentions and thoughts of peace, he was also, I would later discover, the inspiration behind a really quite bizarre form of torture undertaken by U.S. forces in Iraq in May 2003. This torture did not take place in the Abu Ghraib prison, where naked Iraqi detainees were forced to masturbate and to simulate oral sex with one another. It occurred instead inside a shipping container behind a disused railway station in the small town of al-Qā'im, on the Syrian border. It was really just as horrific, in its own way, as the Abu Ghraib atrocities, but because no photographs were taken, and because it involved Barney, the Purple Dinosaur, it wasn't greeted with the same blanket coverage or universal revulsion.

All these things, and the goat staring, and much more besides were inspired by a lieutenant colonel whose name is Jim Channon.

3. THE FIRST EARTH BATTALION

It was a Saturday morning in winter and Lieutenant Colonel Jim Channon (retired) was strolling through the grounds of his vast estate—it stretches across much of a Hawaiian hilltop—yelling above the wind, "Welcome to my secret garden, my eco-homestead. Fresh strawberry? There's nothing like eating something that's just been alive. If the ships stop coming, if history disappears and the world crushes us to death, I shall feed myself. I invite the wind! The wind will come if you ask it to. Do you believe that? Come to my banyan tree. Come this way!"

"Coming!" I said.

The banyan tree was split down the middle, and a crooked cobblestone path wove its way through the roots.

"If you want to pass through these gates," said Jim, "you must be part mystic and part visionary and therefore able to create your best shopping list. So welcome to my sanctuary, where I mend my wounds and dream my dreams about better service."

"Why are you so unlike my mental picture of a lieutenant colonel in the U.S. army?" I asked.

Jim thought about this. He ran his hand through his long, silvery hair. Then he said, "Because you haven't met many of us."

This is Jim now. But it wasn't Jim in Vietnam. Photographs of him back then show a clean-cut young man in military uniform, wearing a badge in the shape of a rifle surrounded by a wreath. Jim still has the badge. He showed it to me.

"What does it mean?" I asked him.

"It means thirty days under combat conditions," he said.

Then he paused and pointed at his badge and said, *"This is real stuff."*

Jim can remember exactly how it all began, the one precise moment that sparked the whole thing off. It was his first day in combat in Vietnam, and he found himself flying in one of four hundred helicopters thundering above the Song Dong Nai River, toward a place known to him as War Zone D. They landed among the bodies of the Americans who had failed to capture War Zone D four days earlier.

"The soldiers," said Jim, "had been cooked in the sun and laid out like a wall."

Jim smelled the bodies, and in that instant his sense of smell shut off. He regained it some weeks later.

An American soldier to Jim's right jumped out of his helicopter and immediately began firing wildly. Jim shouted at him to stop but the soldier couldn't hear him. So Jim leaped on him and wrestled him to the ground.

Jesus, thought Jim.

And then a sniper fired a single shot from somewhere into Jim's platoon.

Everyone just stood there. The sniper fired again, and the Americans started running toward the one and only palm tree in sight. Jim was running so fast that he skidded face first into it. He heard someone behind him shout, "VC in black pajamas, one hundred meters."

About twenty seconds later, Jim thought, Why is nobody shooting? What are they waiting for? They can't be waiting for me to instruct them to shoot, can they?

"TAKE HIM OUT!" screamed Jim.

And so the soldiers started shooting, and when it was over a small team walked forward to find the body. But, for all the gunfire, they had failed to hit the sniper.

How had that happened?

Then a soldier yelled, "It's a woman!"

Oh shit, thought Jim. How do we deal with this?

Moments later, the sniper killed one of Jim's soldiers with a bullet through his lungs. His name was Private First Class Shaw.

"In Vietnam," said Jim, "I felt like tire rubber. The politicians just waved me off. I had to write the letters to the mothers and the fathers of the soldiers who were killed in my unit."

And when he got home to America it was his job to drive out into the countryside to meet these parents and give them citations and the personal belongings of their dead children. It was during these long drives that Jim replayed in his mind the moments that had led to the death of Private First Class Shaw.

Jim had yelled for his soldiers to kill the sniper, and they had all, as one, and with every shot, fired high.

"This came to be understood as a common reaction when fresh soldiers fire on humans," Jim said. "It is not a natural thing to shoot people."

(What Jim had seen tallied with studies conducted after the Second World War by military historian General S. L. A. Marshall. He interviewed thousands of American infantrymen and concluded that only 15 to 20 percent of them had actually shot to kill. The rest had fired high or not fired at all, busying themselves however else they could.

And 98 percent of the soldiers who *did* shoot to kill were later found to have been deeply traumatized by their actions. The other 2 percent were diagnosed as "aggressive psychopathic personalities," who basically didn't mind killing people under any circumstances, at home or abroad.

The conclusion—in the words of Lieutenant Colonel Dave Grossman of the Killology Research Group—was that "there is something about continuous, inescapable combat which will drive 98 percent of all men insane, and the other 2 percent were crazy when they got there.")

For a while after Vietnam, Jim suffered from depression; he found he couldn't watch his daughter being born. He couldn't see anything that reminded him of pain. The nurses in the hospital thought he was crazy because this kind of thing hadn't been explained in the media. It was heartbreaking for Jim to realize that Private First Class Shaw had died because his fellow soldiers were instinctively guileless and kindhearted, and not the killing machines the army wanted them to be.

Jim took me into his house. It looked as though it belonged to some benevolent wizard from a fantasy novel; it was full of Buddhist art, paintings of all-seeing eyes atop pyramids, and so on.

"The kind of person attracted to military service has a great deal of difficulty being . . . cunning. We suffered in Vietnam from not being cunning. We just presented ourselves in our righteousness and we got our butts shot off. You might get some cunning out of other agencies in the American government, but you'd have a hard time finding it in the army."

And so it was, in 1977, that Jim wrote to Lieutenant General Walter T. Kerwin, the vice chief of staff for the army, at the Pentagon. He wrote that he wanted the army to learn how to be more cunning. He wanted to go on a fact-finding mission. He didn't know where. But he wanted to be taught cunning. The Pentagon agreed to pay Jim's salary and expenses for the duration of the journey. And Jim got into his car and began to drive.

Steven Halpern is the composer of a series of meditation and subliminal CDs, sold over the Internet, with titles like *Achieving Your Ideal Weight* ("Play this program during mealtimes. You chew your food slowly. You love and accept your body fully"); *Nurturing Your Inner Child* ("You release any resentment or hurt toward your parents for not meeting your needs"); and *Enhancing Intimacy* ("Your body knows just where to touch me. You love holding and cuddling me").

"For over twenty-five years," reads Steven's web site, "his music has touched the lives of millions, and is used in homes, yoga and massage centers, hospices, and innovative business offices worldwide."

It was at the beginning of Steven's career, in 1978, at a new-age conference in California, that he met Jim Channon. Jim said he wanted somehow to use Steven's music to make the American soldier more peaceful; he also hoped to deploy Steven's music in the battlefield to make the enemy feel more peaceful too.

Steven's immediate thought was, *I don't want to be on a list.*

"Sometimes you end up on a list, you see?" said Steven. "They monitor your activities. Who was this guy? Was he posing as someone who wanted to learn the good things, but was planning to use them against me?"

I was struck by how vividly Steven recalled his encounter with Jim. That, Steven explained, is because people who work in the ambient-music field don't get approached by the military all that often. Plus, Jim seemed to walk the walk in terms of inner peace. Jim was very charismatic. And anyway, Steven added, these were paranoid times. "We'd just come out of Vietnam," he said. "It turned out that some of the most violent antiwar agitators were double agents. It was the same in the UFO community."

"The UFO community?" I said. "Why would government spies want to infiltrate that?"

"Oh, Jon," said Steven. "Don't be naive."

"Why, though?" I asked.

"Everyone kept tabs on everyone," said Steven. "It got so paranoid that UFO speakers would start by asking all the government spies to stand up and identify themselves. The more you know, the more you don't know, see? Anyway, there was a lot of paranoia. And then some guy came over and said he was from the military and he wanted to learn about my music, and that was Jim Channon."

"Why do you think he approached you in particular?" I asked.

"Someone once said that my music allows people to have a spiritual experience without *naming* it," Steven replied. "I think that was it. He said he needed to convince the higher-up military brass, the top ranks. These are people who had never known a meditative state. I think he wanted to get them into it without *naming* it."

"Or maybe he wanted to hypnotize his leaders with subliminal sounds," I said.

"Maybe so," said Steven. "They're very powerful things."

Steven told me a little about the power of subliminal sounds. One time, he said, an American evangelical church blasted the congregation with silent sounds during the hymns. At the end of the service, they found their donations had tripled.

"Tactical advantage, you see?" said Steven. "You want to know why evangelical churches are making so much money while regular churches are failing? Maybe that's your answer."

And recently, he added, he visited a friend's office. "As soon as I walked in there, I felt irritated. I said, 'Your office is making me feel irritated.' He said, 'That's my new subliminal peak-efficiency tape.' I said, 'Well, take it off.'" Steven paused. "I spotted it right away," he said, "because I'm attuned. But most people aren't."

Steven told Jim Channon about the power of subliminal sounds too, and Jim thanked him and left. They never met again.

"This was twenty-five years ago," said Steven. "But I remember it as if it were yesterday. Jim seemed such a gentle soul." Steven fell silent for a moment. Then he said, "You know what? Now that I think back, I'm not sure I ever asked him what he planned to do with all that information."

Almost all the people Jim visited during his two-year journey were, like Steven Halpern, Californians. Jim dropped in on 150 new-age organizations in all, such as the Biofeedback Center of Berkeley; the Integral Chuan Institute ("Just as the bud of a flower contains within it the innate form of the perfect flower, so do we all contain within ourselves the innate form of our own perfection"); Fat Liberation ("You CAN Lose Weight!"); Beyond Jogging; and, in Maine, the Gentle Wind World Healing Organization ("If you attended school in America or a country with similar education practices before the age of ten to twelve years, you suffered severe forms of mental and emotional damage. . . . The Gentle Wind healing technology can help").

Gentle Wind presumably offered Jim, as they have offered all who've passed through their gates, their Healing Instruments, the magic ingredients of which have always been a closely guarded secret, although a clue offered by the company is that they are derived "from the Spirit World, not the human world." Imagine something that looks like a largish bar of white hand soap painted to look like a computer circuit board. That is Gentle Wind's Healing Bar 1.3, "Requested donation $7600." Although pricey, it "represents the new leading edge of the healing technology, significantly surpassing the Rainbow Puck III and IV" and includes "well over

6–60 MHz minimum of temporal shifting combined with millions of predefined etheric modifications."

Gentle Wind's publicity material assures potential purchasers and company recruits that "There Are No Messiahs Here . . . NO MESSIAHS at the Gentle Wind Project. Please do not waste your time looking for any. There are none here."

Nonetheless, some former members have alleged to me that Gentle Wind's chief guru, John Miller, has over the past few years ordered his entire staff to go on the Atkins diet and wear only beige, and that the mysterious spirit-world ingredient incorporated into their Healing Bars is actually group sex. The alleged scenario is apparently something like this: John Miller sidles up to a female staff member and says— and I am paraphrasing, based on allegations made to me by former members of the group—"Congratulations. You have been selected by the spirit world to take part in our top secret energy work. Don't tell your husband because he wouldn't understand the energy work."

She is then led into John Miller's bedroom, has sex with him and various female selectees, and then, the moment it is over, John Miller says, "Quick. Build a Healing Bar." Gentle Wind and its leaders, including Miller, are contesting these allegations and, in 2004, launched a lawsuit against the former members who made them.

One Gentle Wind customer review—from a couple in Bristol—reads, "We have noticed remarkable improvements to our cat Moya, who virtually changed overnight from a neurotic timid rescue cat into a friendly confident adventurer after we used one of the Gentle Wind healing instruments on her."

Another customer, however, has noted, "At first I was pleased that the device did have a noticeable effect on my aura [but when I turned it over to the label marked Tranquillity] it left me feeling inwardly unresponsive to the experience. To cut a long story short, I have been using Equilibria's Universal Harmonizers for the past five months instead and now feel very much myself again."

Gentle Wind says that over 6 million people in more than 150 countries have used their products. They also told me that they don't remember meeting Jim, and perhaps it was a different Gentle Wind he came across during his Pentagon-funded odyssey. They could be right, but I have not managed to find another Gentle Wind operating within the new-age or human-potential movement at that time.

Jim Channon couldn't remember much about Gentle Wind either, although the group must have made an impact on him because he gave them a special mention in the confidential report he later prepared for the Pentagon.

Jim went through Reichian rebirthing, primal arm wrestling, which was regular arm wrestling accompanied by guttural screaming, and naked hot-tub encounter sessions at the Esalen Institute for the Advancement of Human Potential, in Big Sur, where he was counseled by Esalen's founder, Michael Murphy, the man credited with inventing the new-age movement. At no time did Jim reveal to the therapists and gurus he met how he imagined their techniques might be adapted to teach the American soldier to be more cunning.

"It is often ten years," Jim wrote in his diary at the time, "before the values developed in Los Angeles find themselves into rural Arkansas. What is developing today on the Coast will be the national value set ten years from now."

This is how Jim visualized the America of the 1980s: The government would no longer have an "exploitative view of natural resources." Instead, its emphasis would be on "conservation and ecological sanity." The economic system would cease to "promote consumption at all costs." It would be neither aggressive nor competitive. This, Jim prophesied, was the new value system, poised to sweep America.

Jim needed to believe that all this would happen. He was working for what his then chief of staff, General Edward Meyer, had called a "hollow army." This was a term Meyer had coined to describe the military state of mind, post-Vietnam. It wasn't only individual veterans who were suffering from postcombat depression and posttraumatic stress. The army, as an entity, was traumatized and melancholic and suffering from a crippling inferiority complex. Budgets were being slashed everywhere. The draft had been abolished, and the army was not an enticing career option for young Americans. Things were in a really bad state. Jim saw himself as a potential new-age phoenix, rising from the ashes to bring joy and hope to the army, and to the country he loved so much.

"It is America's role," wrote Jim, "to lead the world to paradise."

Jim returned from his journey in 1979 and wrote a confidential paper for his superiors. The first line read, "The U.S. army doesn't really have any serious alternative than to be wonderful."

A disclaimer at the bottom read, "[This] does not comprise an official position by the military as of now."

This was Jim Channon's *First Earth Battalion Operations Manual.*

The manual was a 125-page mixture of drawings and graphs and maps and polemical essays and point-by-point redesigns of every aspect of military life. In Jim Channon's First Earth Battalion, the new battlefield uniform would include pouches for ginseng regulators, divining tools, foodstuffs to enhance night vision, and a loudspeaker that would automatically emit "indigenous music and words of peace."

Soldiers would carry with them into hostile countries "symbolic animals" such as baby lambs. These would be cradled in the soldiers' arms. The soldiers would learn to greet people with "sparkly eyes." Then they would gently place the lambs on the ground and give the enemy "an automatic hug."

There was, Jim accepted, a possibility that these measures might not be enough to pacify an enemy. In that eventuality, the loudspeakers attached to the uniforms would be switched to broadcast "discordant sounds." Bigger loudspeakers would be mounted on military vehicles, each playing acid rock music out of sync with the others to confuse the enemy.

In case all that didn't work, a new type of weaponry—nonlethal or "psychoelectronic" weapons—would be developed, including a machine that could direct positive energy into hostile crowds.

If all else failed, lethal weapons would be used, although "no Earth soldier shall be denied the kingdom of heaven

peace technology

indigenous music and words of peace

country flags + spiritual symbols

the battalion carries the symbols and sounds of peace

symbolic flowers

symbolic animal

THE FIRST EARTH BATTALION

because he or she is used as an instrument of indiscriminate war."

Back on base, robes and hoods would be worn for the mandatory First Earth Battalion rituals. The misogynistic and aggressive old chants ("I don't know but I've been told, Eskimo pussy is mighty cold . . .") would be phased out and replaced by a new one: "Om."

Military marching bands would learn how to become more like traveling minstrels. "Singing and dancing" and "the elimination of the desire for lust" would be as important a part of training as martial arts.

"A Warrior Monk is one who has no dependence on lust," wrote Jim. "A Warrior Monk is one who has no dependence on status. This regimen is not meant to produce puritanical fanatics but is clearly designed to exclude the soldier of fortune."

(The above portion of the manual was, presumably, disregarded by Michael Echanis, who became America's most famous soldier of fortune in the years between allegedly staring a goat to death at Fort Bragg and dying under mysterious circumstances in Nicaragua.)

First Earth Battalion trainees would learn to fast for a week drinking only juice and then eat only nuts and grains for a month. They would:

> fall in love with everyone, sense plant auras, organize a tree plant with kids, attain the power to pass through objects such as walls, bend metal with their minds, walk on fire, calculate faster than a computer, stop their own hearts with no ill effects, see into the future, have out-of-body experiences, live off nature for twenty days, be 90%+ a vegetarian, have the ability to massage and cleanse the colon, stop using mindless clichés, stay out alone at night, and be able to hear and see other people's thoughts.

Now all Jim had to do was sell these ideas to the military. I think Jim Channon is a wealthy man. He certainly owns pretty much an entire Hawaiian hillside, with an amphitheater, a village worth of outbuildings, yurts, and gazebos. Nowadays he does for corporations what he did for the army: he makes their employees believe they can walk through walls and change the world, and he does it by making those things sound ordinary.

"Do you honestly believe," I asked Jim at one point during our day together, "that somebody can reach such a high level of warrior monkdom that they can actually become invisible and walk through walls?"

Jim shrugged.

"Women have been known to lift up an automobile single-handed when their child is under it," he said. "Why not expect the same from a Warrior Monk?"

Jim told me—just as he had told his commanding officers back in 1979—that "warrior monk" might sound like a crazy new military prototype, but was it any more crazy than the old prototypes, like cowboy, or football player?

"A Warrior Monk," said Jim, "is someone who has the presence of the monk, the service and the dedication of the monk and the absolute skill and precision of the warrior."

He had told his commanders this at the officers' club in Fort Knox in the spring of 1979. He had arrived there a few hours earlier and had dragged in as many potted plants as he could find around the base. He arranged them into a circle, a "pseudo-forest." In the center of the circle he lit a single candle.

When the commanders arrived, he said to them, "To begin the ceremony, gentlemen, we're going to do a mantra. Take a deep breath and as you let it out sound *eeeeeeee.*"

Jim told me, "At this point, they laughed. A few of them chuckled, a little bit embarrassed. So I was able to say, '*Excuse* me! You've been given a set of instructions and I expect them to be carried out at high level.' See? Tapping right into the military mind-set. Second time we did it, the place became unified."

And then Jim began his speech. He said, "Gentlemen, it is

a great honor to have you in this place of sanctuary where we can mend our wounds and dream our dreams of better service. Together, with all the other armies of the world, we will turn this place around, and a new civilization can be born that does not know boundary lines but knows better how to live in the garden and knows that we are one thought away from paradise."

The commanders were not laughing anymore. In fact, Jim found that some were almost in tears. Like Jim, they had been crushed by their experiences in Vietnam. Jim was speaking to four-star generals and major generals and brigadier generals and colonels—"the top people"—and he had them captivated. In fact, one colonel present, Mike Malone, was so moved that he leaped to his feet and yelled, "I am mullet man!"

Noticing the perplexed expressions on the faces of his fellow military commanders, he elucidated. "I push the cause of the mullet because he is a low-class fish. He is simple. He is honest. He moves around in great formations and columns. He does damn near all the work. But he is also *noble*. He is like another noble thing I once loved, called 'soldier.'"

Jim continued his speech: "The only time that rose-colored glasses don't work is when you take them off," he said. "So join me in this vision of being *all* that we can be, for this is the place where the First Earth Battalion begins. This is the place where you have the right to think the unthinkable, to dream the impossible. You know we're here to create the most powerful set of tools for the individual and his team, for that is the difference between where the American soldier is today and where he needs to be to survive on the battle-field of the future."

"You know what this story is about?" Jim asked me, in his garden in Hawaii. "It's the story of the creativity of an institution you would expect to be the *last* to open the door to the greater realities. Because you know what happened next?"

"What?" I asked.

"I was immediately appointed commander of the First Earth Battalion."

The disclaimer at the bottom of Jim's *Operations Manual* had read that this was not the official position of the United States military. Nonetheless, within weeks of its publication, soldiers throughout the army began seriously to try to implement his ideas.

Somewhere in a strip mall in the heart of Silicon Valley is a building that looks like a long-abandoned and entirely undistinguished warehouse. Nonetheless, busloads of tourists turn up from time to time to photograph the exterior because this is the building where Silicon Valley began. It started life as an apricot-storage warehouse, but then Professor William Shockley moved in and coinvented the transistor and grew silicon crystals in the back room and won the Nobel Prize for his work in 1956.

By the late 1970s this building—391 San Antonio Road—had a new owner, Dr. Jim Hardt. He was just as much of a pioneer in his field as Shockley, just as much of a visionary, but his science was, and remains, somewhat weirder.

Dr. Hardt still works there, charging civilians $14,000 for a week-long brain-training retreat—"Mention the keyword 'hemi-coherent' and get a $500 discount!" says the publicity pack—in a series of tiny offices at the back. They are dark, lit only in fluorescent purple, the clocks have no hands, and the place reminded me a little of Disney World's Twilight Zone Tower of Terror ride.

I had come to believe that Michael Echanis was not, after all, the fabled goat starer. I had decided that Glenn Wheaton had been mistaken, beguiled by the Echanis legend, and that it was another Jedi Warrior altogether. Perhaps Dr. Hardt might be able to provide the answer, for it was he who retuned the brains of the Jedi Warriors in the late 1970s, and took them to a level of spiritual enlightenment within which staring a goat to death was, apparently, possible.

Dr. Hardt sat me down and he told me the story of his "fascinating, yet somewhat melodramatic" adventures with Special Forces.

It all began with a visit from a colonel named John Alexander, who turned up one day at Jim Hardt's door with a few other military men. Colonel Alexander had headhunted Dr. Hardt, having been deeply moved by Jim Channon's *First Earth Battalion Operations Manual.* He wanted to know if Dr. Hardt could really turn ordinary soldiers into advanced Zen masters in just seven days, and give them the power of telepathy simply by plugging them into his brain machine.

Dr. Hardt said it was indeed true, and so the quest to create a supersoldier, a soldier with supernatural powers, was set into motion right there in that building in Silicon Valley.

The colonel told Jim Hardt that Special Forces had, ever

since the publication of Jim's manual, invited one peak-performance guru after another from the new-age and human-potential movements of California to lecture the soldiers on how to be more attuned with their inner spirits, and so on, but it had not been a success. The gurus had routinely been greeted with boos, catcalls, and theatrical yawns by Special Forces.

Now, Colonel Alexander wanted to know, would Dr. Hardt be willing to give it a try? Would he bring his portable brain-training machine to Fort Bragg?

Jim Hardt showed me the machine. You strap electrodes onto your head and your alpha waves are fed into a computer. Knobs are tweaked and your alpha waves are attuned. When this has been achieved your IQ is boosted by twelve points and you effortlessly reach a spiritual level usually attainable only through a lifetime's diligent study of Zen techniques. If two people are strapped to the machine simultaneously, they get to read each other's minds.

Dr. Hardt explained all this to Colonel Alexander, and he offered to give him a demonstration, but Colonel Alexander declined. He said there was a lot of classified military information stored in his brain and he couldn't risk telepathically revealing it to Dr. Hardt.

Dr. Hardt said he understood.

Colonel Alexander felt obliged to tell Dr. Hardt that Special Forces were really quite hostile to the whole idea, which they considered mumbo-jumbo. They would in fact be "uncontrollable" and refuse to "sit still and listen."

In that case, Dr. Hardt replied, he would accept the challenge only if the soldiers were first sent on a month-long meditation retreat.

"Well," Dr. Hardt said to me now. "First of all, they wouldn't call it a meditation *retreat,* because retreat is a no-no word in the army. So it was called a meditation *encampment.* And it was *hugely* unsuccessful."

"How come?" I asked.

"The soldiers actually *brawled* with each other in the meditation setting," he said. "They brawled out of boredom."

And so, by the time Dr. Hardt arrived at Fort Bragg, Special Forces were still "extremely hostile," blaming Dr. Hardt for their month's enforced meditation, which they had considered "nonsense" and "a waste of time."

The small, thin, and delicate Dr. Hardt anxiously surveyed the hostile soldiers, then he gently strapped the electrodes onto their heads, and onto his head too. He switched on the alpha-wave brain-training computer, and the tuning began.

"And then suddenly," said Jim Hardt, "a tear came out of my eye and it rolled down my face and it splashed onto my tie."

A tear almost formed in his eye now, as he recalled this moment of emotional telepathy.

"So I picked up my tie, it was still wet, and I said, 'I telepathically know that somebody in this room is experiencing sadness.' And I slammed my hand down on the table and I said, 'We are not leaving this room until whoever it is owns up to it.' Well. Two minutes of total silence. And then this hardbitten colonel raised his hand and he said, 'That was probably me.'"

And then the colonel told Jim Hardt, and his fellow Special Forces soldiers, the story of his sadness.

This colonel had sung in his glee club at college. He had sung folk and choral music and, as his brain was being tuned, his mind filled with the memories of his glee club days some twenty years earlier.

"He experienced such joy in that," said Jim Hardt. "But then he went straight from college to officer training school and he made an intellectual decision to give up on joy. He decided on graduating from college that joy had no role in the life of an army officer and so he consciously and willfully, *click,* turned joy off. Now it was twenty years later, and he came upon the realization that it wasn't necessary. He had lived twenty years without joy. And it wasn't necessary."

On day two of the brain tuning, the soldiers strapped the electrodes to their heads once again.

"And this time," said Jim Hardt, "both of my eyes were like faucets. And I took my tie and I wrung it. That's how soaked it was with my tears, and so again I said, 'Who *is* it? Who is experiencing sadness?' And again it was two minutes before the *same guy* raised his hand and this time he recounted a story that he had lived through."

It was the Tet offensive in 1968. The colonel was in a small forward firebase up by the demilitarized zone when the Vietcong attacked.

"And this colonel single-handedly saved their little firebase from being overrun," said Jim Hardt, "and the way he did this was by running the machine gun all night long. And then, when dawn came, he looked out at the piles of bleeding, dying bodies that he had caused, and he had feelings that are larger than one heart can encompass."

At the end of day three of the brain tuning, Jim Hardt studied the alpha-wave computer printouts, and he saw something that amazed him.

"In one of the soldiers," he said, "I saw a pattern of brain waves which is found only in people who have experience of seeing angels. We call it 'perception of astral plane beings,' beings that are discorporate but have a luminous body. So I was sitting across the desk from this soldier who had been trained to kill, and I asked him, in a very calm voice, 'Do you talk to beings that other people don't see?'

"And he spun back in his chair. He almost tipped over. It was like I had hit him with a two-by-four! And he was all nervous and alarmed and his breathing was heavy, and he looked left and right, to ensure that nobody else was in the room. Then he leaned forward and admitted it, 'Yes.' He had a martial arts spirit guide who would appear to him alone. And he had only told his best buddy about this, and he had sworn that he would cut his throat if his friend breathed a word of this to anyone."

And that was the end of the story. That was all Dr. Hardt could tell me. He left Fort Bragg, never returned, and said he didn't know which, if any, of the Jedi soldiers whose brains he tuned had gone on to stare a goat to death.

"Nonlethals only!" yells the evil medical researcher Glenn Talbot. "I repeat, *nonlethals only*! I must have a sample of him. Hit him with the foam!"

In the underground Atheon military base, hidden beneath a disused cinema in a desert somewhere, the Incredible Hulk has escaped and is destroying all in his path. The soldiers do what Glenn Talbot has ordered. They take up position and spray the Hulk with Sticky Foam, which expands and hardens the moment it hits his body. The foam succeeds where all previous weapons have failed. The Hulk is stopped in his tracks. He struggles, roaring, against the foam, to no avail.

"So long, big boy . . ." snarls Glenn Talbot. He shoots the Hulk in the chest with some kind of handheld missile launcher. This is a mistake. It makes the Hulk angrier—so angry, in fact, that he summons enough power to break through the foam and continue his rampage.

This foam is not an invention of the writers of the *Hulk* movie. It is the invention of Colonel John Alexander, the same man who recruited Dr. Jim Hardt to retune the brains of the Jedi Warriors. Colonel Alexander developed the Sticky Foam as a result of reading Jim's *First Earth Battalion Operations Manual.*

The army leaders present at Fort Knox back in 1979 had been so taken with Jim's speech that they offered him the opportunity to create and command a real First Earth Battalion. But he turned them down. Jim had higher ambitions than that. He was rational enough to realize that walking through walls, sensing plant auras, and melting the hearts of the enemy with baby lambs were good ideas on paper, but weren't, necessarily, achievable skills in real life.

Jim's superiors were literal-minded men (hence General Stubblebine's many determined efforts to walk through his wall), but Jim's real vision was more nuanced. He wanted his fellow soldiers to find a higher spiritual plane by reaching for the impossible. Had he accepted the offer of leading a real

First Earth Battalion, his superiors would have demanded measurable results. They would have wanted to see Jim's soldiers *demonstrably* stopping their own hearts with no ill effects, and when they failed, the unit would most probably have been shut down, in ignominy, without anyone really knowing it had existed.

This was not what Jim had in mind. He wanted his ideas to float out there and take root wherever fate decreed. The First Earth Battalion would exist wherever someone read the manual and became inspired to implement its contents however he chose. Jim imagined it would be assimilated into the fabric of the army so successfully that the soldiers of the future would act on it without knowing anything about its fantastic provenance. And so it was that Sticky Foam became an early, real-life, First Earth Battalion weapon.

The foam has had a rocky history. In Somalia in February 1995, United Nations peacekeeping forces were attempting to hand out food when the crowd began to riot. U.S. Marines were brought in to calm things down and aid in the UN's withdrawal.

"Use the Sticky Foam!" ordered the commanding officer. And the Marines did. They sprayed the foam not into the crowd, but in front of it so that it would harden and produce an instant wall between the rioters and the food. The Somali crowd paused, looked at the bubbling, expanding, hardening, custardlike substance, waited for it to solidify, climbed over it, and carried on rioting. All this occurred in front of the TV cameras. That night, news broadcasts across America ran the footage alongside a clip from *Ghostbusters* in which Bill Murray was slimed.

(One of the deployers of the Sticky Foam in Somalia—Commander Sid Heal—later warned me against portraying the incident as an unmitigated disaster. He said they had hoped it would take the rioters twenty minutes to figure out how to scale the foam, but instead it took them five minutes, and so the worst you could say was that it was a three-quarters disaster. It was, however, the first and last time that the foam was deployed in a combat situation.)

Unperturbed by the Somali incident, the U.S. penal authorities introduced Sticky Foam into prisons in the late 1990s to subdue violent inmates before they were transported elsewhere. The practice was quickly discontinued, however, because it was impossible to move the foamed prisoners from their cells once they'd been immobilized. They were just stuck there.

But now, unexpectedly, the foam is enjoying a renaissance. Bottles of the stuff were taken to Iraq in 2003. The idea was that once U.S. troops found the weapons of mass destruction, Sticky Foam would be sprayed all over them. But the weapons of mass destruction were never found, and so the foam remained in its bottles.

Of all Jim's ideas, the most fruitful was his insistence that military operatives and scientists should journey to the wildest corners of their imaginations, unafraid to appear harebrained and half-baked in their pursuit of a new kind of weapon, something cunning and big-hearted and non-lethal.

The foam is one of hundreds of similar inventions mentioned in a leaked 2002 air force report—*Non-Lethal Weapons: Terms and References*—which comprehensively details the latest endeavors in this field. There are a number of acoustic weapons: the Blast Wave Projector, the Curdler

Unit, and the low-frequency Infrasound, which, according to the leaked report, "easily penetrates most buildings and vehicles" and creates "nausea, loss of bowels, disorientation, vomiting, potential internal organ damage or death." (Jim Channon's successors seem more laissez-faire about their definition of the term *non-lethal* than he was.) Then there are the Race-Specific Stink Bomb and the Chameleon Camouflage Suit, neither of which has gotten off the ground yet, because nobody can work out how to invent them.

There is a special pheromone that "can be used to mark target individuals and then release bees to attack them." There's the Electric Glove, the Electric Police Jacket, "which jolts anyone who touches it," the Net Gun, and the Electric Net Gun, which is the same as the Net Gun but "will release an electric shock if the target tries to struggle." There are all manner of holograms, including the Death Hologram—"used to scare a target individual to death. Example, a drug lord with a weak heart sees the ghost of his dead rival appearing at his bedside and dies of fright"—and the Prophet Hologram, "the projection of the image of an ancient god over an enemy capital whose public communications have been seized and used against it in a massive psychological operation."

The First Earth Battalion's Colonel John Alexander is named as a coauthor of the report. He lives in the suburbs of Las Vegas, in a large house filled with Buddhist and aboriginal art and military awards. There were also, I noticed, a number of books written by Uri Geller on his shelf.

"Do you know Uri Geller?" I asked him.

"Oh yes," he said. "We're great friends. We used to have metal-bending parties together."

Colonel Alexander has been a special adviser to the Pentagon, the CIA, the Los Alamos National Laboratory, and NATO. He is also one of Al Gore's oldest friends. He is not completely retired from the military. A week after I met him, he flew to Afghanistan for four months to act as a "special adviser." When I asked him who he was advising and on what, he wouldn't tell me.

For much of the afternoon, instead, John reminisced about the First Earth Battalion. His face broke into a broad smile when he recalled the secret late-night rituals that he and some fellow colonels would enact on military bases, after reading Jim's manual.

"Big bonfires!" he said. "And guys with snakes on their heads!"

He laughed.

"Have you heard of Ron?" I asked him.

"Ron?" said Colonel Alexander.

"Ron who reactivated Uri," I said.

Colonel Alexander fell silent. I waited for him to respond. After about thirty seconds, I realized that he wasn't going to say another word until I asked him a different question. So I did.

"So did Michael Echanis really kill a goat just by staring at it?" I asked.

"Michael Echanis?" he said. He looked perplexed. "I think you're talking about Guy Savelli."

"Guy Savelli?" I asked.

"Yes," said the colonel. "The man who killed the goat was definitely Guy Savelli."

4. INTO THE HEART OF THE GOAT

The Savelli Dance and Martial Arts Studio stands around the corner from a Red Lobster, a TGI Friday's, a Burger King, and a Texaco garage, in the suburbs of Cleveland, Ohio. The sign on the door advertises lessons in ballet, tap, jazz, hip-hop, aerobatics, pointe, kickboxing, and self-defense.

I had telephoned Guy Savelli a few weeks earlier. I told him who I was and asked if he might describe the work he had undertaken inside Goat Lab. Colonel Alexander had told me that Guy was a civilian. He was under no military contract. So it seemed possible that he might talk. But instead there was a very long silence.

"Who are you?" he finally asked.

I told him again. Then I heard a profoundly sad sigh. It was something more than "Oh no, not a journalist." It sounded almost like a howl against the inescapable and unjust forces of destiny.

"Have I called at a bad time?" I asked him.

"No."

"So *were* you at Goat Lab?" I asked.

"Yes." He sighed again. "And yes, I *did* drop a goat when I was there."

"I don't suppose you still practice the technique?" I asked him.

"Yes, I do," he said.

Guy fell silent again. And then he said—and his voice sounded sorrowful and distressed—"Last week I killed my hamster."

"Just by staring at it?" I asked.

"Yes," confirmed Guy.

Guy was a little more relaxed in the flesh, but not much. We met in the foyer of his dance studio. He is a grandfather now, but still jumpy and full of energy, moving around the room as if possessed. He was surrounded by some of his children and grandchildren, and half a dozen of his *Kun Tao* students stood anxiously around the edges of the studio. Something was up, that was clear, but I didn't know what.

"So you did this to your hamster?" I asked Guy.

"Huh?" he said.

"Hamsters," I said, suddenly unsure of myself.

"Yes," he said. "They . . ." A look of bewilderment crossed his face. "When I do it," he said, "the hamsters *die*."

"Really?" I asked.

"Hamsters drive me nuts," said Guy. He began talking very fast. "They just go around and around. I wanted to stop it from going around and around. I thought, *I'm going to make it sick so it'll burrow under the sawdust or something*."

"But instead you made it die?"

"I've got it on tape!" said Guy. "I taped it. You can watch

the tape." He paused. "I had a guy take care of the hamster every night."

"What do you mean?" I asked.

"Feed it. Water it."

"So you knew it was a healthy hamster," I said.

"Yes," said Guy.

"And then you started staring," I said.

"Three days," sighed Guy.

"You must hate hamsters," I said.

"It's not that I *want* to do that to hamsters," Guy explained. "But supposedly, if you're a *master*, you should be able to do that kind of stuff. Is life just a punch and a kick and that's it? Or is there more to it than that?"

Guy jumped in his car and went off to find his home video of the hamster being stared to death. While he was gone, his children, Bradley and Juliette, set up a video camera and began to film me.

"Why are you doing that?" I asked them.

There was a silence.

"Ask Dad," said Juliette.

Guy returned an hour later. He was carrying a sheaf of papers and photographs along with a couple of video cassettes.

"Oh, I see Bradley has set up the camera," he said. "Don't worry about that! We film everything. You don't mind, do you?"

Guy put the tape into the VCR, and he and I began to watch.

The video showed two hamsters in a cage. Guy explained to me that he was staring at one, trying to make it sick and visibly paranoid about its wheel, while the other was to remain throughout an unstared-at control hamster. Twenty minutes passed.

"I've never known a hamster," I said, "so I—"

"Bradley!" interrupted Guy. "You ever own a hamster?"

"Yes," replied Bradley.

"You ever see one do *that* before?"

Bradley came into the room and watched the video for a moment.

"Never," he said.

"Look at the way it's glaring at the wheel!" said Guy.

The target hamster did indeed seem suddenly mistrustful of its wheel. It sat at the far end of the cage, looking at it warily.

"Usually that hamster *loves* its wheel," explained Guy.

"It does seem odd," I said, "although I have to say that emotions such as circumspection and wariness are not that easy to discern in hamsters."

"Yeah, yeah," said Guy.

"There will be some people reading this who own hamsters," I said.

"Good," said Guy. "Then they'll know how *rare* that is. Your hamster people will know that."

"My hamster-owning readers," I agreed, "will know whether or not this is aberrant behavior. . . . He's down!" I said.

The hamster had fallen. Its legs were in the air.

"I'm accomplishing the task I wanted to do," said Guy. "Look! The other one has run right over it! He's right on top of the other hamster! That's bizarre! That's kind of nuts, isn't it? He's not moving! I'm accomplishing my task right there."

The other hamster fell over.

58

"You've dropped *both* hamsters!" I said.

"No, the other one has just fallen over," explained Guy.

"Okay," I said.

There was a silence.

"Is he dead now?" I asked.

"It gets more bizarre in a minute," said Guy. He seemed to be dodging the question. "*Now!* It's more bizarre now!"

The hamster was motionless. And it remained that way—utterly immobile—for fifteen minutes. Then it shook itself down and began eating again.

And then the tape ended.

"Guy," I said. "I don't know what to make of this. The hamster *did* seem to be behaving unusually in comparison with the control hamster, but on the other hand it definitely didn't die. I thought you said I was going to watch it die."

There was a short silence.

"My wife said, 'No,'" he explained. "Back at the house. She said, 'You don't know if this guy's a bleeding-heart liberal.' She said, 'Don't show him the hamster *dying*. Don't show him that. Show him the tape where the hamster acts bizarre instead.'"

Guy told me that what I had just seen were the edited highlights of two continuous days of staring. It was on the third day, Guy said, that the hamster dropped dead.

"I am a ghost," said Guy.

We were in the foyer of his dance studio, standing underneath the bulletin board. It was covered with mementos of Savelli family successes. Jennifer Savelli, Guy's daughter, danced with Richard Gere in *Chicago*. She danced in the seventy-fifth Academy Awards. But there was nothing much on the wall about Guy—no newspaper cuttings or anything like that.

"You would never have known about me if Colonel Alexander hadn't told you my name," he said.

It was true. All I could find about Guy in the newspapers was the odd notice from the Cleveland Plain Dealer concerning awards won by his students in local tournaments. This other side of his life was entirely unchronicled.

Guy riffled through the papers and the photographs.

"Look!" he said. "Look at this!"

He handed me a diagram.

"Guy," I said. "Is this Goat Lab?"

"Yes," said Guy.

Bradley silently filmed me studying the Goat Lab diagram.

Then Guy dropped the papers and the photographs. They scattered all over the floor. We both bent down to pick the stuff up.

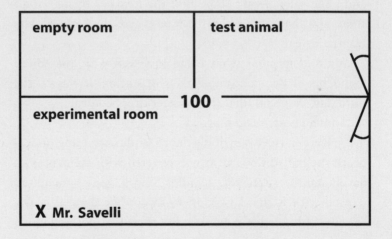

"Oh," murmured Guy. "You weren't supposed to see *that*."

I quickly looked over. In the moments before Guy hid it between some documents, I caught a glimpse of what I wasn't supposed to see.

"Bloody hell," I said.

"Right," said Guy.

It was a blurry snapshot of a soldier crouched in a frosty field next to a fence. The photograph appeared to have captured the soldier in the act of karate chopping a goat to death.

"Jesus," I said.

"You really weren't supposed to see that," said Guy.

Guy's story began with a telephone call he received, out of the blue, in the summer of 1983.

"Mr. Savelli?" said a voice. "I'm phoning from Special Forces."

It was Colonel Alexander.

Guy was not a military man. Why were they calling him? The colonel explained that since their last martial arts teacher, Michael Echanis, had died in Nicaragua back in 1978, Special Forces had basically stopped incorporating those kinds of techniques into their training programs down at Fort Bragg, but they were ready to give them another try. They had chosen him, he explained, because the branch of martial arts he practices—Kun Tao—has a uniquely mystical dimension. Guy teaches his students that "only with total integration of the mental, physical, and spiritual can one hope to come away unscathed. Our intention is to teach this integrated way and show others how to have exceptional paranormal results that are usually associated with fables for the young."

The colonel asked Guy if he could come down to Fort Bragg for a week or so, to test the waters. Could he step into Michael Echanis's shoes? Guy said he'd give it a try.

On the first day, Guy taught the soldiers how to break slabs of concrete with their bare hands, how to withstand being whacked on the back of the neck with a thick metal rod, and how to make a person forget what he is about to say.

"How do you make a person forget what he's about to say?" I asked Guy.

"Easy," he said. "You just do this—" Guy scrunched up his face and yelled, "Noooooooo!"

"Really?" I asked.

"You ever play pool and you miss your shot and you want your opponent to miss his shot and you go, 'Noooooooo'? And then they miss *their* shot! It's the same thing."

"Is it all in the tone of voice?" I asked.

"You say it *inside* your head," said Guy, exasperated. "You get that *feeling* inside of you."

And so it was, on the evening of the first day, that Special Forces mentioned to Guy that they had goats. Guy said he couldn't remember who steered the conversation then, but he did recall, at some point during the evening, announcing, "Let's give it a try."

"So the next morning," Guy said, "they got a goat, they set it up, and we started."

As Guy recounted this story, the atmosphere inside the dance studio remained apprehensive. Bradley continued to film me. From time to time, when we made small talk about holidays or the weather, I could see what a lovely family the Savellis were—close-knit, tough, and intelligent. But whenever we returned to the subject of goats, the mood instantly hardened.

It turns out that the goat Guy stared at had not been debleated or shot in the leg. Guy had said he wanted a normal, healthy goat, so that's what they gave him. It was herded into a small room that was empty but for a soldier with a video camera. Guy knelt on the floor in another room.

And he began to get that feeling inside him.

"I pictured a golden road going up into the sky," he said. "And the Lord was there, and I walked into His arms, and I got a chill, and I *knew* it was right. I wanted to find a way to knock that goat down. We have this picture of St. Michael the Archangel with a sword. So I thought about that. I thought about St. Michael with this sword going . . ."

Guy mimed the action of St. Michael violently thrusting his sword downward into a goat.

" . . . Through the goat and . . ."

Guy smacked his hands together.

" . . . Knocking it down to the ground. Inside of me I couldn't even breathe. I was going . . ."

Guy mimed struggling for breath.

"And you *believe* it," he said. "You *believe* it. And after about fifteen minutes I said, 'Lenny, you better go see. I don't know for sure.'"

Lenny from Special Forces disappeared into the room where the goat was. He came back and announced, with surprise and solemnity, "The goat is down."

"And that was it?" I asked.

"That was it," said Guy. "It lay there for a while, and then it got up again."

"Is that the end of the story?" I asked.

"No," said Guy sadly. "I wish it was. But the next day they wanted me to do it *again*. But this time they wanted me to *kill* the goat. They said, *'Kill the goat!'*"

He fell silent, as if to say, *See what I had to deal with?*

"Why *kill* the goat?" I asked.

"Military people," sighed Guy. "I guess they thought you could . . . whatever . . ."

"Okay," I said.

So on day three a new experiment was devised. Guy told Special Forces to round up thirty goats.

"Thirty goats," he told them. "Put numbers on them. I'll pick a number. I'll drop the goat."

On this occasion, Special Forces stationed armed guards all around the Goat Lab perimeter. There had been no such security the day before, presumably because they hadn't really imagined that a goat would topple. But this time, Guy told me, the mood was far more somber. Thirty goats, all with numbers strapped to their backs, were herded inside. Guy randomly chose number 16. And he began.

But this time, he said, he just couldn't concentrate. Whenever he pictured himself walking into the arms of the Lord, his meditation was disrupted by the memory of a Special Forces soldier yelling *"Kill the goat"* at him. He got as far as picturing St. Michael the Archangel, but just as he was about to thrust his sword downward, the cry of *"Kill the goat"* again interrupted the psychic path between Guy and the animal.

"I was just so pissed off," Guy said. "Anyway, when

Lenny went next door to look it turned out that number *seventeen* had dropped dead."

"Collateral damage?" I said.

"Right," said Guy.

And that, he said, was the end of his story.

Except for one last thing. Ten years later, Guy said, three Special Forces soldiers covertly traveled to Cleveland from Fort Bragg, having heard on the grapevine that Guy had once successfully stared a goat to death on their base. They wanted to know if the rumor was true. They wanted to see it for themselves. They wanted Guy to kill a goat for them.

But Guy said no. He had killed enough goats for one lifetime. He was beginning to feel the dark forces of karma descend upon him. So he offered them a compromise. He would teach the soldiers how to do it for themselves. So the Special Forces men arranged to meet Guy at the office of a local veterinarian who had agreed to provide a goat and an ECG machine.

"You brought a veterinarian into this?" I asked, surprised.

"Yeah, the guy was a friend of mine," said Guy.

"And he provided the goat?"

"Yeah."

"What about the Hippocratic oath?" I asked.

"What?" said Guy, a little crossly.

"I'm just surprised that a civilian veterinary surgeon would provide a healthy goat so some soldiers could try to stare it to death."

But Guy shrugged and said I didn't have to take his word for it and he put a cassette into his VCR and pressed Play.

And I saw that it was true. A strange tableau flickered

onto the screen, the opening scene of a goat snuff movie. A goat was strapped to an ECG machine. The veterinarian was nowhere to be seen, but the office was clearly a vet's office, with certificates on the wall and various animal medical-type implements scattered around. Two soldiers in combat fatigues sat on plastic chairs taking notes. The goat bleated. The soldiers continued to take notes. The goat bleated again. The ECG machine bleeped. The soldiers took more notes. Guy nudged me in the ribs.

"Ha!" he said. "Dear me!" He chuckled. "This isn't even the best part yet."

"Is somebody staring at the goat?" I asked.

"Yeah," he said. "That guy."

"Which one?" I asked. "That one or that one?"

"Neither," said Guy. *"That* one."

Guy pointed to a corner of the screen at something I hadn't noticed—the shoe of a third man, just off camera.

For another ten minutes the bleating, bleeping, and note taking continued on the VCR.

"Am I going to see some kind of physical response on the goat's part?" I asked Guy.

"It's happening now!" said Guy. "Look at the machine. The heart rate was around the midsixties. Now it's dropped to fifty-five."

"Oh," I said.

The video ended. Guy turned off the TV. He seemed a little annoyed at my disappointed tone of voice.

"Let me get this completely clear," I said. "What I just saw was level one."

"Right," said Guy. "The goat was attached to the life force of the man off screen."

"And if you carry that further," I said, "to level two, the goat will drop, or fall, or topple over, or tumble."

"Yes," said Guy.

"So the hamster was level two?" I asked.

"Right," said Guy.

"And if you go even further than that, the goat or the hamster will die."

"Yes." Guy paused. "But level one is high!" he said. "Hey! Level one is high!"

"Does it hurt to be the recipient of level one?" I asked.

"No," said Guy.

"Guy," I said recklessly, "will you stare at me?"

There was a silence.

"Not this time," said Guy softly. "When you come back I will. This time, my wife said, 'No.' She said, 'You don't know this person.' Which is true. She said, 'Don't do anything this time.' She said I have too much trust in everybody. And I do. I do, I do."

My day with the Savellis was over, so I thanked them and got ready to leave. It was then that Guy gently tapped my shoulder and said, "There's something you should know."

"Mmm?" I said.

And he told me.

And then it all made sense—the profound sigh I had heard down the phone when I'd first called Guy, the shock on everyone's face when I saw the snapshot of the goat being karate chopped to death in a frozen field, Bradley's constant

67

filming of me. Guy told me everything, and when he finished explaining, I said, "Oh my God."

Guy nodded.

"Bloody hell," I said. "Really?"

"Really," said Guy.

"Jesus," I said.

5. HOMELAND SECURITY

Six years before Major General Albert Stubblebine III failed to walk through his office wall in Arlington, Virginia, his office didn't exist. There was no INSCOM—the army's Intelligence and Security Command. There were just military intelligence units scattered haphazardly around the world. It was—according to the author Richard Koster, who served with the 470th Counter Intelligence Corps Detachment in Panama during the pre-Stubblebine days—chaos.

"In the late 1950s," Koster told me when I phoned him to ask about life in military intelligence before Stubblebine, "there were frantic calls from commander to commander. 'We need to greatly expand military intelligence. We want you to release X number of officers. We need a colonel, three majors, six captains, and fifteen lieutenants to be immediately reassigned to military intelligence.' So what do you do if you get a call like that? You think, *Ha! Let's give them all our bindlestiffs and stumblebums*. So they did. And that's who went to army intelligence, more or less globally."

"What was it like in Panama before General Stubblebine?" I asked him.

"This was not a tight ship," he said. "We had a riot one year here in Panama City. My colonel came running up to me. 'Where is this riot?' I said, 'It's right in front of the Legislative Palace.' He said, 'Where's that?' I said, 'Go to the Tivoli Hotel. You'll see it out the balcony.' He looked at me like I was Einstein because I had this . . . *knowledge.*"

In the late 1970s a brigadier general named William Royla was given the job of tidying up the whole mess. He was to form a kind of CIA for the army; it would be called INSCOM. And in 1981 General Stubblebine, who had been deeply moved by Jim Channon's *First Earth Battalion Operations Manual* and was filled with the conviction that America, the great superpower, needed to be defended by people who actually had superpowers, was appointed its commander.

Stubblebine was a West Point man with a master's degree in chemical engineering from Columbia. He learned about the First Earth Battalion when he was stationed at the Army's Intelligence School in Arizona. It was his friend and subordinate, Colonel John Alexander, the inventor of Sticky Foam, who first drew his attention to it.

Now, General Stubblebine was determined to turn his sixteen thousand troops into a new army, an army of soldiers who could bend metal with their minds and pass through objects and consequently never have to go through the chaotic trauma of a war like Vietnam again. Who would want to mess with an army like that?

Plus, Stubblebine's tenure as commander of military intelligence coincided with huge slashes in his budget. These were the post-Vietnam "draw down" days, and the Pentagon wanted their soldiers to achieve more with less money. Learning how to walk through walls was an ambitious but inexpensive enterprise.

And so it was that Jim Channon's madcap vision, triggered by his postcombat depression, found its way into the highest levels of the United States military.

Twenty years later, in room 403 of the Tarrytown Hilton, in a suburb of New York City, just as General Stubblebine had finished describing his failed attempts to walk through his wall, he glanced out the window.

"A cloud," he said.

The three of us—the general, his second wife, Rima, and I—rose from our chairs.

"Jesus, Jon, I don't know," said the general. "I've never done one that big."

All day we had been waiting for the right sort of cloud to come along, a cumulus, in fact, so he could show me that he could burst it just by staring at it. Of all his powers this was, he said, the easiest to demonstrate.

"Anyone can see it," he had promised me, "and anyone can do it."

"Right in the notch, way over where the pine trees are," said Rima. "Do *that* one."

"Let me see," said the general.

He stood very still and began to stare up at the sky.

"Are you trying to burst that one over *there?*" I asked. "Isn't it too far away?"

General Stubblebine looked at me as if I were nuts.

"They're *all* far away," he said.

"Over there!" said Rima.

I darted my eyes back and forth across the sky, trying to work out which cloud the general was trying to burst.

"It's *gone!*" said Rima.

"The cloud," confirmed the general, "appears to have gone."

We sat back down. Then the general said he wasn't sure. The clouds had been moving so fast, he said, it wasn't possible to conclude 100 percent that he had caused the disappearance. It might have been just meteorology.

"Hard to tell," he said, "who was doing what to whom."

Sometimes on long car journeys, he said, Rima would drive and he would make the clouds go away, and if it was a puffy cloud alone in a blue sky, it was unequivocal. He would stare: the cloud would burst. But this wasn't one of those moments.

In 1983, two years into his tenure as the commander of military intelligence, General Stubblebine's pursuit of an indisputable miracle became an urgent one. He needed something to satisfy his commanding officers in the Pentagon, and he needed it fast, because his job was in jeopardy.

General Stubblebine was confounded by his continual failure to walk through his wall. What was wrong with him that he couldn't do it? Perhaps there was simply too much in his in-box to give it the required level of concentration. General Manuel Noriega, principally, was causing him significant trouble in Panama. Noriega had been on the U.S. intelligence payroll since the 1970s—since CIA director George H. W. Bush had authorized his recruitment—but now he was out of control.

General Stubblebine's CIA counterparts had been using

Panama's network of hidden airstrips to transport guns to the Contras in Nicaragua. Once the weapons had been delivered, the planes returned to Panama to refuel for their journey back to the United States. Noriega seized the opportunity to fill them with cocaine. And so it was that the CIA became implicated in Noriega's cocaine racket. This awkward alliance was making both sides paranoid, and when General Stubblebine visited Panama, he discovered to his fury that Noriega had his hotel room bugged.

It was at this point that the battle between the two generals—Noriega and Stubblebine—shifted into the supernatural. Noriega took to tying black ribbons around his ankles and placing little scraps of paper in his shoes with names written on them to protect him against spells cast by his enemies. He was possibly walking around Panama City with the word *Stubblebine* secreted inside his shoe at the very moment that the general was trying to walk through his wall. How could General Stubblebine concentrate on passing through objects with that sort of craziness going on around him?

General Stubblebine countered by setting his psychic spies on Noriega. This was the Fort Meade team, who worked out of a condemned clapboard building down a wooded track in Maryland and who, as a result of not officially existing, had no coffee budget, a fact that they had come to resent. They were also going stir-crazy. Their offices were claustrophobic, and many of them didn't much like one another to begin with. One, a major named Ed Dames, had taken to psychic spying on the Loch Ness monster during the fallow months, when there wasn't much official military psychic work. He determined that it was a dinosaur's ghost. This finding irritated some of the others, who considered it unscientific and frankly implausible. Another psychic spy, David Morehouse, was soon to check himself into a psychiatric hospital as a result of an excess of psychic spying.

They couldn't get their back door open. It had been locked and painted shut dozens of times over the years. Nobody knew where the key was. During one particularly hot day they began almost to faint in there, and so the talk got around to whether they should kick the door open and get a breeze going through.

"We can't," said Lyn Buchanan. "We don't *exist*. If we kick it open, nobody will come and fix it."

(It was Lyn Buchanan who recounted this story to me, when I met him in the summer of 2003 at a hotel in Las Vegas.)

"Leave it to me," said psychic spy Joe McMoneagle. He disappeared and returned twenty minutes later with a detailed and psychically divined sketch of the missing key. Joe McMoneagle then drove into town to a local locksmith, got the key made from the sketch, returned to the unit, unlocked the back door, and pried through the paint.

"Oh, Joe's good," said Lyn Buchanan. "Joe is very good."

I visited Joe McMoneagle a few months later. He lives in Virginia now. I mentioned Lyn Buchanan's story about the key. After I told him what Lyn had said, Joe smiled somewhat guiltily.

"I, uh, actually picked the lock," he admitted.

He explained that Lyn had seemed so bedazzled, and it had given the flagging morale of the psychic spies such a boost, that he hadn't had the heart to inform them of the fact that the door was opened using nonpsychic means.

Working conditions at Fort Meade were so grim that a conspiracy theory began to flourish within its condemned walls. There they were, hitherto ordinary soldiers who had been handpicked and initiated into a fabulously secret military psychic elite, which turned out to be utterly humdrum. Lyn Buchanan and some of his colleagues had consequently come to believe that there must be *another* secret psychic unit, even more deeply embedded, and presumably with more glamorous offices than theirs.

"I got to think that we were there in order to be caught," Lyn said when I met him in Las Vegas.

Lyn is a soft-eyed, folksy-looking man who—for all the dismal working conditions—sees his time in the old unit as the happiest days of his life.

"What do you mean, 'there to be caught'?" I asked him.

"You know," said Lyn. "If the *National Enquirer* ever got wind of it, the army could have said to them, 'Yes, we *do* have a secret psychic unit. Here they are.'"

Hang the psychics out to dry—postulated Lyn with some bitterness—so that the other psychics, whoever they were, would be left in peace to continue their even more secret work.

So in the summer of 1983, when General Stubblebine asked the team to divine in which room of a particular villa in Panama City Noriega was staying, and what Noriega was thinking about while he was there, they sprang into action, delighted for some distraction.

General Stubblebine simultaneously ordered a team of

nonpsychic spies to rent an apartment down the road from Noriega's villa. The timing was critical. The moment the Fort Meade psychics delivered their divinations, General Stubblebine phoned the nonpsychics in Panama and ordered them to climb over the wall, get inside the villa, and plant bugs in Noriega's rooms. Unfortunately, two of Noriega's guard dogs were alerted during the covert raid, and the nonpsychics were chased back over the wall.

General Noriega responded to this assault by placing a huge amulet around his neck and driving to a nearby beach where his personal sorcerer, a Brazilian named Ivan Trilha, erected an illuminated cross to ward off American intelligence operatives.

General Stubblebine had his adversaries at home too. His superior officer, General John Adams Wickham, the army's chief of staff, was not a fan of the paranormal. General Stubblebine had attempted to captivate him at a high-level black-tie party in a Washington hotel by producing from his tuxedo pocket a piece of bent cutlery, but General Wickham recoiled, horrified.

The reason General Wickham felt the way he did about bent cutlery can be found in Deuteronomy, chapter 18, verses 10–11:

"There shall not be found among you *any one* that maketh his son or his daughter to pass through the fire, or that useth divination . . . or an enchanter, or a witch, or a charmer, or a consulter with familiar spirits, or a wizard, or a necromancer."

General Wickham believed, and in fact told colleagues, that Satan had somehow taken over General Stubblebine's soul. It was Satan, not General Stubblebine, who had bent the fork.

In later White House administrations, including that of George W. Bush, General Wickham has continued to command respect. In his autobiography, Colin Powell twice refers to him as "my mentor," and in June 2002 he received George W. Bush's American Inspirations Award for his work as part of the Presidential Prayer Team, a 3-million-strong community of Americans who log on to presidentialprayerteam.org every week to be told what to pray for:

> Pray for the ongoing efforts in the war on terror, that the President and all his intelligence sources will obtain the most helpful information in safeguarding America. Pray for them to have godly wisdom in the manner in which they handle each bit of information. Pray for the effectiveness of a new fingerprinting initiative that will screen foreign travelers entering America. Pray for the strong relationship between Mr. Bush and Mr. Blair. Pray that the President will continue to be guided by the Lord in his deliberations with the U.K.

And so on. General Stubblebine might have suggested to General Wickham that prayer groups were not dissimilar to spoon-bending-type initiatives, both being attempts to harness the power of the mind to influence things from afar, but the general's unassailable enemy regarding this logic was Deuteronomy, chapter 18, verses 10–11.

Funnily enough, and unknown to General Wickham, General Stubblebine had in fact undertaken every one of the above abominations before the Lord during his tenure as head of army intelligence, with the exception of making his son or daughter pass through fire, although he *had* fire-walked himself in the mountains of Virginia, under the tutelage of the self-help guru Anthony Robbins.

General Wickham's hard-line interpretation of Deuteronomy was making General Stubblebine's position untenable, hence his urgent need to come up with an indisputable miracle. Back home in Arlington, his late-night attempts at levitation met with no success. The general put this failure, too, down to his ever-burgeoning in-box, which is why he eventually flew to Fort Bragg in an attempt to persuade Special Forces to burst the hearts of animals just by staring at them. If he didn't have the time to perfect these powers, perhaps they might.

It is hard to predict whether General Stubblebine might have found a kindred spirit in his commander in chief, President Reagan. The president seemed to have a foot in both camps. His chief of staff, Donald Regan, wrote in his memoirs that "virtually every major move and decision the Reagans made during my time as White House chief of staff was cleared in advance with a woman in San Francisco who drew up horoscopes to make certain that the planets were in a favorable alignment for the enterprise."

This woman, whose name was Joan Quigley, fixed the exact time when the president would sign the Intermediate Nuclear Forces treaty in 1987. Joan Quigley now goes by the presumably unauthorized title Presidential Astrologer Joan Quigley.

But the president also shared, with his friend General

Wickham, an abiding respect for the fundamentals of the Bible. When the states of Arkansas and Louisiana passed a law stating that creationism be taught in public schools, the president cheered the initiative, announcing, "Religious America is awakening!"

When I telephoned General Wickham to ask for his account of that black-tie party, he said he remembered it well. It was a big dinner at a place called Quarters One. He couldn't recall specifically blaming Satan. But, yes, he had recoiled, he said, because as a Christian you have to accept that the supernatural is alive, and it sometimes manifests itself in eerie ways. But General Stubblebine was, broadly speaking, "one of the good guys."

"I became actually kinda intrigued," he told me.

General Stubblebine had spotted a flash of curiosity cross General Wickham's face at the party, and he recognized that this could be a watershed moment in military history. If he could only beguile his famously Christian chief of staff by performing an on-the-spot paranormal demonstration, might this be the moment when the supernatural began its journey toward official recognition by the U.S. army?

This is why General Stubblebine seized the opportunity to say to General Wickham, "I can do it for you *now* if you like. I can bend a spoon for you right now, if you like."

And this, General Wickham told me, was General Stubblebine's error.

"I didn't want him to bend a spoon in the middle of a *party,*" he said. "It was an inappropriate place to do it."

It was exactly this sort of overenthusiasm that led to General Stubblebine's enforced early retirement.

But the supernatural war against Manuel Noriega did not end with General Stubblebine's departure. Five years later, in December 1989, the United States launched Operation Just Cause to depose Noriega and put him on trial for cocaine smuggling. But when American troops arrived in Panama, they discovered that Noriega had gone into hiding.

An agency within the U.S. government (Sergeant Lyn Buchanan told me he couldn't remember which it was, and anyway, he said, the information was probably still classified) called up the psychic spies. Where was Noriega? Lyn Buchanan sat inside the clapboard building in Fort Meade, put himself into a trance, and received "a powerful impulse regarding Noriega's location."

"Ask Kristy McNichol," he kept writing on a piece of paper. "Ask Kristy McNichol."

Sergeant Buchanan was certain that the TV actress Kristy McNichol, who appeared in *Starsky & Hutch*, the ABC miniseries *Family, The Bionic Woman,* and *The Love Boat II,* held the key to the whereabouts of General Noriega. At that time, in December 1989, Kristy McNichol had just recorded the CBS special *Candid Camera! The First 40 Years,* had a guest role in *Murder, She Wrote,* and had starred in the erotic thriller *Two Moon Junction.*

"Ask Kristy McNichol," Lyn continually wrote, in his trance state.

Lyn Buchanan stopped at this point and said he didn't know whether anyone had acted on his divination. The way the secret psychic unit was structured, he explained, meant that once his divinations had been passed upward, he was rarely given feedback about what happened next. He had no idea if the authorities subsequently contacted Kristy McNichol.

So I attempted to ask her myself. I e-mailed her to inquire whether by chance she had known where General Manuel Noriega was holed up in December 1989. In addition, was I the first person to have approached her about this matter, or had others, perhaps U.S. intelligence operatives, contacted her in the past?

I never got a reply.

For everyday agnostics, it is not easy to accept the idea that our leaders, and the leaders of our enemies, sometimes seem to believe that the business of managing world affairs should be carried out within both standard and supernatural dimensions.

Over the course of a year or two I contacted everyone I could find who had met Jim Channon during his late-1970s Californian odyssey. One of them was Stuart Heller. Stuart had been introduced to Jim by their mutual friend Marilyn Ferguson—the renowned author of *The Aquarian Conspiracy*. Stuart told me that Jim was "just marvelous."

These days, Stuart teaches business executives the art of stress control. He visits Apple and AT&T and the World Bank and NASA and coaches their managers in how to remain centered and tranquil amid the workplace hurly-burly. He is one of scores of similar gurus who travel from business to business throughout the Western world, fulfilling Jim's 1979 prophecy that "what is developing today on the Coast will be the national value set ten years from now."

At one point during my conversation with Stuart I happened to ask him if he knew anyone who was the living embodiment of the First Earth Battalion. Stuart instantly replied, "Bert Rodriguez."

"Bert Rodriguez?" I said.

"He's a martial arts guy down in Florida," Stuart said. "My younger brother is one of his students. I've never met anybody like Bert. His gym is always full of ex-military guys, ex-Special Forces. Spooks. And in the middle there's my skinny little brother."

I typed Bert Rodriguez's name into a search engine and my screen filled with a picture of an intense-looking shaved-headed Cuban with a black mustache, frozen in the act of slamming a huge and sweaty man into the wall of his gym—the US 1 Fitness Center in Dania Beach, Florida.

"Bert once got my brother to lie on the floor," Stuart said, "and he put a cucumber on his chest, and he blindfolded himself and *wham*! He sliced the cucumber in half with a samurai sword. Didn't cut my brother at all. *Blindfolded*!"

"Bloody hell," I said.

"Bert's one of the most spiritual guys I've ever met," said Stuart. "No. Spiritual is the wrong word. He's occultic. He's like a walking embodiment of death. He can stop you at a distance. He can influence physical events just with his mind. If he catches your attention he can stop you without touching you."

Stuart paused.

"But he doesn't talk like this. He's the most First Earth Battalion guy I know but he's incapable of verbalizing it. He's a street fighter from Cuba. With Bert it's just instinctive.

But *everyone* can see it. That's why people come and train under him."

In April 2001, Bert Rodriguez took on a new student. His name was Ziad Jarrah. Ziad just turned up at the US 1 Fitness Center one day and said he had heard that Bert was good. Why Ziad chose Bert, of all the martial arts instructors scattered around the Florida shoreline, is a matter of speculation. Maybe Bert's uniquely occultic reputation preceded him, or perhaps it was Bert's military connections. Plus, Bert had once taught the head of security for a Saudi prince. Maybe that was it.

Ziad told Bert that he was a businessman who traveled a great deal and he wanted to learn how to defend himself if a group attacked him.

"I liked Ziad a lot," Bert Rodriguez said when I called him. "He was very humble, very quiet. He was in good shape. Very diligent."

"What did you teach him?" I asked.

"The choke hold," said Bert. "You use it to put someone to sleep or kill them. I taught him the choke hold and the kamikaze spirit. You need a code you'd die for, a do or die desire. And that's what gives you the sixth sense, the ability to see *into* the opponent and know if he's bluffing. Yeah. I taught him the choke hold and the kamikaze spirit. Ziad was a soccer player. I'd much rather have a soccer player beside me in a fight than a black belt in Tae Kwon Do. The soccer player can dodge and dive."

There was a silence.

"Ziad was like Luke Skywalker," said Bert. "You know when Luke walks the invisible path? You have to believe it's there. And if you do believe it, it *is* there. Yeah. Ziad believed it. He was like Luke Skywalker."

Bert trained Ziad for six months. He liked him, sympathized with his tough upbringing in Lebanon. He gave Ziad copies of three of his knife-fighting training manuals, and Ziad passed them on to a friend of his, Marwan al-Shehhi, who was staying up the road in room 12 of the Panther Motel and Apartments in Deerfield Beach, Florida.

We know this because when Marwan al-Shehhi checked out of the Panther Motel on September 10, 2001, he left behind a flight manual for a Boeing 757, a knife, a black canvas bag, an English–German dictionary, and three martial arts manuals written by Bert Rodriguez, the man Stuart Heller had called "the most First Earth Battalion guy I know."

Marwan al-Shehhi was twenty-three years old when he checked out of the Panther Motel, flew to Boston, changed planes, took control of United Airlines flight 175, and crashed it into the south tower of the World Trade Center.

Ziad Jarrah was twenty-six when he took control of United Airlines flight 93, which came down in a field in Pennsylvania on its way to Washington, D.C.

"You know what?" said Bert. "I think Ziad's role was to be the hijacker with brains. He'd hang back to ensure that the job was done properly, that the takeover of the plane was completed." Bert paused. "If you love a son and he becomes a mass murderer, you don't stop loving your son, do you?"

Guy Savelli's role in the War on Terror began when half-a-dozen strangers, within days of one another, contacted him via e-mail and telephone in the winter of 2003. They asked him if he had the power to psychically kill goats. Guy was bewildered. He did not go around publicizing this. Who were these men? How did they know about the goats? He feigned a casual tone of voice and said, "Sure I can."

Then he immediately phoned Special Forces.

Everyone who had contacted him, he told them, was *Muslim,* with the possible exception of some British guy (me). The others were certainly e-mailing from Muslim countries, axis-of-evil countries, in fact. This had never happened to Guy before. Might they be al-Qaeda? Might they be bin Laden operatives hoping to learn how to stare people to death? Was this the start of a whole new paranormal subdivision of al-Qaeda?

Special Forces instructed Guy to meet me, because in all probability I was al-Qaeda too.

"Be careful what you say to him," they advised.

Special Forces had even—I was startled to learn—been on the phone to Guy the very morning I had visited him. While I was getting coffee at the Red Lobster, they had phoned Guy and said, "Has he turned up yet? Be careful. And film him. Get him on tape. We want to know who these people are. . . ."

I'm not sure at what stage during the day we spent together Guy decided that I wasn't an Islamic terrorist. Perhaps it was when I discovered that his daughter danced with Richard Gere in the movie *Chicago* and I screeched, "Catherine Zeta-Jones was *brilliant* in it!"

Even a deep-cover al-Qaeda terrorist wouldn't think to go *that* fey.

I do know that throughout our entire hamster conversation, Guy was still convinced that I wasn't an actual journalist. When I spoke of my "hamster-owning readers" Guy had glanced dubiously at me because he believed I had no readers at all, and would be reporting the events of the day not to the public but to a terrorist cell.

That was the reason—Guy explained—why such a panicked scramble had ensued when I spotted the snapshot of the soldier karate chopping a goat to death. It was no ordinary karate chop, Guy revealed. It was the death touch.

"The death touch?" I asked.

Guy told me about the death touch. It was, he said, the fabled Dim Mak, also known as the quivering palm. The death touch is a very light strike. The goat is far from whacked. Its skin isn't broken. There isn't even a bruise. The goat will then stand there with a dazed expression on its face for about a day, before it suddenly topples over, dead.

"Imagine if al-Qaeda had that kind of power," Guy said. "Staring is one thing. The death touch is quite another. That's why we were all so freaked when you saw the picture. We still didn't know if you were al-Qaeda."

And so it was that Guy's life had taken a strange new twist. Was he to be a dance and martial arts instructor by day and a covert agent infiltrating a hitherto unknown paranormal unit of al-Qaeda by night?

Over the next few weeks Guy and I kept in touch.

"I met with *another* department," he told me during one call.

"Homeland Security?" I asked.

"I can't tell you *that,*" said Guy. "But they're sure one of the guys who contacted me is al-Qaeda. They're *sure* of it."

"How do they know?" I asked.

"The name checks out," said Guy. "The phone number too. The phone number is on a *list.*"

"What did the intelligence people say to you?" I asked.

"They said, 'Yeah, yeah. He's one of the guys for sure.'"

"Al-Qaeda?"

"Al-Qaeda," said Guy.

"Are you *bait*?" I asked.

"That's what it looks like," said Guy. "It's getting kinda hairy here."

"You're bait," I said.

"I'll tell you, Jon," said Guy, "these intelligence people see me as a dog. A *dog*! I said to them, 'I have a family.' 'Yeah yeah,' they said. 'Having a family is very, very nice.' We're really expendable. I'm going to end up hanging from a lamp-post. A fucking *lamppost.*"

At this point I heard Guy's wife say, "Very fucking funny."

"Hang on," said Guy.

Guy and his wife had a muffled conversation.

"My wife says I shouldn't be talking like this on the phone," he said. "I'm hanging up now."

"Keep me informed!" I said.

And Guy did. As the various schemes to ensnare the possible al-Qaeda paranormal subdivision changed, Guy kept me informed of the developments. Plan A was for Guy to invite these people to America. Then the intelligence people had a change of heart, telling Guy, "We don't want them *here.*"

The much riskier plan B was for Guy to travel to *their* country. He would teach them a relatively benign psychic power and report back everything he saw and heard.

Guy told them, "No fucking way."

Plan C was for Guy to meet them on neutral ground—maybe London. Or France. Plan C suited both camps and seemed the most likely to proceed.

"I would fucking love to have you there," Guy said.

Guy sent me a scrap of an e-mail that, he told me, was "absolutely, positively" written by an al-Qaeda operative. It read:

Dear sir Savelli,

I hope you are fine and fit. I am bussy in my champion ship my champion ship is going success ful. Sir Savelli, please tell me if I apply to affiliation in your Federation so what is a prosiger please tell me a detal.

And that was it. It seemed that one of two scenarios was unfolding: Guy was either in the middle of a sensational sting operation, or a hapless young martial arts enthusiast who only wanted to join Guy's federation was about to be shipped off to Guantanamo Bay. All we could do was wait.

6. PRIVATIZATION

This has so far been a story about secret things undertaken clandestinely inside military bases in the United States. From time to time tangible results of these covert endeavors have made their way into everyday life, but always far removed from their supernatural roots. Nobody who came into contact with Colonel Alexander's Sticky Foam, for example— not the prisoners who were glued to their cells by it, not the TV crews who filmed its partially disastrous deployment in Somalia, not even, I would guess, the soldiers who carried it into Iraq in the hope of spraying it all over the WMDs—was aware that it was the product of a paranormal initiative from the late 1970s.

All of a sudden, though, in 1995, a palpable chunk of the craziness leaked from the military community into the civilian world. The man who did the leaking was an errant prodigy of General Stubblebine.

This is what happened.

As a child, growing up in the 1970s, Prudence Calabrese loved watching *Doctor Who* and science documentaries. She lived in a run-down mansion in New England. When her parents went out on Saturday nights, the children would whip out their homemade Ouija board and try to contact the ghost of the previous owner who had apparently hanged herself in the barn as a result of being alcoholic and unpopular with the neighbors. They held pajama-party séances.

"We wanted to have unusual experiences," Prudence told me as we sat at her kitchen table in Carlsbad, near San Diego. "We would all get together and light candles and turn the lights down and try to make a table rise just by touching it."

"Did it ever rise?" I asked her.

"Well, yes," said Prudence. "But we were kids. Looking back, I'm not sure if everybody just added a little bit of effort and that made it rise."

"With your knees?" I said.

"Yeah," said Prudence. "Hard to tell."

Sometimes Prudence and her friends would run outside and try to spot UFOs. They thought they saw one once.

Prudence went to the local university but she got pregnant when she was eighteen, so she dropped out and began managing a local trailer park with her first husband, Randy. She moonlighted as a dancer in a pig costume at the state fair, went back to college, studied physics, dropped out, had another four children, taught belly dancing to pensioners in Indiana and finally ended up with a new husband named Daniel in an apartment in Atlanta, running a web-site-design business. It was here, in 1995, that Prudence turned on the TV one night. A military man was on the screen.

"What was he saying?" I asked Prudence. "Didn't he say he was a real-life Obi-Wan Kenobi?"

"That's exactly the words he used," said Prudence. "A 'real-life Obi-Wan Kenobi.'"

"Working for the U.S. military?"

"Working for the U.S. military," said Prudence.

"And until that moment nobody even knew that these people existed?" I asked.

"Yeah," said Prudence. "Until that time they had been kept completely secret. He was talking about how he used just his mind to access anything in the whole universe. And how the military used him, and other psychic spies like him, to avert wars and discover secret things about other countries. He said they were called remote viewers. Yeah. According to his story, he was part of a secret team of psychic spies, and he was one of the leaders of the unit. And he just didn't look like what you'd expect. He didn't look like he had super secret powers."

"What did he look like?"

Prudence laughed.

"He was short and scrawny and he had this crazy hairdo from the seventies, and a mustache. And he didn't even look like a military guy, let alone a psychic spy. He just looked like a weird person, a person you'd see on the street."

The man on the TV said he had top-level clearance. He said he knew the exact location of Saddam Hussein and the lost ark of the covenant. Prudence was transfixed. As she watched the TV, her long-forgotten childhood passions came back to her: the Ouija board, *Doctor Who,* the science projects she used to do at school.

"I remembered why I was so excited about science fiction and reading all those stories about psychics and aliens," she said.

Prudence determined in that moment that this was what she wanted to do with her life. She wanted to be like the man on the TV, to know the things he knew, to see the things he could see.

His name was Major Ed Dames.

General Albert Stubblebine had been happy to discuss with me his inability to pass through walls and to levitate, and his apparent failure to interest Special Forces in his animal-heart-bursting initiative. He recounted those incidents to me in a jolly way, even though they can't have been good memories for him. The only time during our meetings that an anguished look crossed his face was when the conversation turned to the subject of his prodigy, Major Ed Dames.

"It bothered me so badly that he talked," he said. "There he was, yap, yap, yap, yap, yap." The general paused. "Yap, yap, yap, yap, yap," he said, sadly. "If anybody should have had a gag put in his mouth it was Ed Dames. He *clearly* was out talking when he should have been listening. Very upsetting, incidentally."

"Why?"

"He'd taken the same oath I took: 'I swear I will not divulge.' But he ran over everyone to talk. He puffed up his chest. 'I was one of them!' He wanted to be king."

Ed Dames had been one of General Stubblebine's personal recruits. When the general took command of the secret psychic unit in 1981, he allowed a bunch of fellow enthusiasts from within the military to join the program. The government's psychic research had, until that time, basically centered on three men: an ex-policeman and building contractor named Pat Price, and two soldiers, Ingo Swann and Joe McMoneagle. These three were regarded by all but the most hardened skeptics to have some kind of unusual gift. (Joe McMoneagle's gift apparently manifested itself after he fell out of a helicopter in Vietnam.)

But General Stubblebine passionately believed the First Earth Battalion doctrine that every human being alive was capable of performing supernatural miracles, so he opened wide the doors of the secret unit, and Ed Dames came in.

As a child, Ed Dames had been a great fan of Bigfoot, UFOs, and sci-fi shows. He had heard rumors about the unit while he was stationed, conventionally, up the road from the psychic spies at Fort Meade, so he petitioned General Stubblebine to let him in. Perhaps this is why the general remains so angry with Ed Dames nine years after Prudence watched him reveal the secrets of the unit on TV that night. Maybe he feels partly responsible for the terrible things—involving Prudence—that happened next.

In 1995, Ed suddenly, and repeatedly, spilled the beans in a big way. He took to appearing on TV shows and radio shows. He didn't mention the goat staring, or the wall walking, or the First Earth Battalion, but he spoke with relish about the secret psychic unit.

But it was the Art Bell show that really turned Ed into a superstar.

Art Bell broadcasts from the very small desert town of Pahrump, Nevada. Pahrump seldom hits the news, although it did once make the headlines for having America's highest suicide rate per capita. Nineteen of Pahrump's thirty thousand townspeople are inclined to kill themselves each year. Pahrump is also home to the world's most famous brothel, the Chicken Ranch, a few dusty streets away from which lies Art Bell's house. This is blue and sprawling and fenced off and surrounded by antennae. Art Bell may be situated in the middle of nowhere, and his show may go out in the dead of night, but he is syndicated on more than five hundred AM stations to an audience of something like 18 million Americans.

At his peak, I have been told, Art Bell had 40 million listeners, many of whom were attracted by the appearance of Ed Dames. Dames became something of a regular fixture on the show. Here is a typical excerpt from one of his appearances in 1995.

ART BELL: If you'll recall, the government, over many years now, has dumped a lot of money and time and effort into remote viewing. So, it's not as crazy as it might seem. I managed to get Major Dames on the line. I know it's very, very late. Major, welcome to the program.

ED DAMES: Thank you, Art.

ART BELL: What can you tell us?

ED DAMES: Well, in addition to our training, and our high-level contracts that we perform for various agencies—tracking terrorists for the government—we have data indicating that human babies will be dying soon, many human babies. . . . It appears there is a bovine AIDS virus developing. This bovine AIDS will become a toxico-

logical insult to human babies and they will die in relatively large numbers.

ART BELL: God. Whew! . . . No escape, huh?

ED DAMES: No, there doesn't appear to be an escape.

ART BELL: Oh, God, this is *horrible* news.

Art Bell has played host to many prophets of doom over
the years, but this one, sensationally, was a major in the
United States Army with top-level security clearance. Ed continued: Yes, millions of American babies were imminently to
develop AIDS from drinking infected cows' milk. This was,
he said, something he had psychically perceived while still in
the army, and he had passed the information to his superiors.

So the highest-ranking military intelligence officers knew
this too.

Art Bell gasped at the revelation that advance knowledge
of this impending cataclysm went to the very top.

Furthermore, Ed said, 300-mph winds were soon to ravage America, wiping out all the wheat, and everyone would
have to stay indoors for pretty much the rest of their lives.

"It was great!" reminisced Prudence at her kitchen table in
San Diego. "These were the glory days of remote viewing.
People were so excited about it. It seemed so fantastic. Ed
Dames immediately became one of Art Bell's very favorite
interviewees ever. He was on *all* the time. He said we were
going to be scorched by this huge solar flare, which was going
to wipe out most of life on Earth. And he said that an incom-
ing comet, Hale-Bopp, was going to drop a plant pathogen."

"Really?" I asked.

"Yeah. He said an alien race had attached a canister to Hale-Bopp and it was going to drop this canister on Earth and some kind of virus was going to come out and eat all the plant life and we'd have to live on earthworms and live underground." Prudence laughed.

"Ed Dames said that?"

"Oh yeah! And he had specific dates for this. He said it was going to happen by February 2000."

We both laughed.

"And what about the bovine AIDS?" I asked.

"Bovine AIDS!" said Prudence. She turned serious. "Mad Cow," she said.

Between 1995 and today, in addition to the bovine AIDS and the 300-mph winds, Major Ed Dames has publicly predicted the following, mostly on the Art Bell show: pregnant Martians living underground in the desert will emerge to steal fertilizer from American companies; AIDS will be found to have originated in dogs, not monkeys; flying fungus from outer-space cylinders will destroy all crops; the existence of Satan, angels, and God will be proved beyond all doubt; and lightning on a golf course in April 1998 would kill President Clinton.

"And mixed up with this," said Prudence, "he talked about his experiences with the military, which made all that wacky stuff seem so much more real and tangible. The government did not dispute that he was a psychic spy; they lauded his efforts; he won *medals*. He was honorably discharged. Everything about him checked out."

"It must have sometimes sounded to some of Art Bell's listeners like they were eavesdropping on top-level meetings inside the Pentagon," I said.

"It sounded so real," said Prudence. "He would talk about how the military had put twenty million dollars of taxpayers' money into the research, so it all made sense."

What Art Bell's listeners didn't know was that Ed Dames was an atypical military psychic spy. Most of Ed's colleagues in the secret unit at Fort Meade spent their time psychically viewing extremely boring things, mostly map coordinates. Ed, meanwhile, was psychically concluding that the Loch Ness monster was the ghost of a dinosaur. Had one of Ed's less-colorful contemporaries chosen to spill the beans instead, and gone on Art Bell to talk about map coordinates, I doubt that the listening millions would have been so spellbound.

Ed's media appearances may have hastened the demise of the secret unit. The CIA officially declassified it and shut it down in 1995. General Stubblebine's foot soldiers had been trying to be psychic for the best part of their careers, and now it was over. After years of living simultaneously in a world where they routinely shot forward and back in time and space—inside Noriega's living room in Panama City one minute, psychically creeping through Saddam Hussein's palaces in Iraq the next—they emerged into perhaps the strangest world of all: the civilian world.

For a while in the mid-1990s it looked like there might be a lot of money to be made. Ed Dames moved to Beverly Hills, where he took high-level meetings with Hollywood executives. He began dealing with Hanna-Barbera, the makers of *Scooby Doo*, about the prospect of transforming himself into a cartoon character for a Saturday-morning kids' show about supersoldiers who used their psychic powers to defeat evildoers. He set up a psychic spying training school, charging students twenty-four hundred dollars for a "highly personalized (one on one), rigorous four-day program."

His company slogan was "Learn Remote Viewing from the Master."

On a Saturday in the summertime, Ed Dames and I roared through Maui in his jeep. (Like Jim Channon, and Sergeant Glenn Wheaton, who first let slip to me that Special Forces had undertaken covert goat-staring activities in Fort Bragg, Ed has set up home in the Hawaiian Islands.)

Ed wore big wraparound sunglasses—his eyes were the only part of his face that looked his age. Ed is fifty-five now, but everything else about him is teenage—his surfer hair, his torn jeans, his manic energy. He held a Starbucks coffee in one hand and steered the jeep with the other.

"Were people in the military cross with you for spilling the beans about the existence of the secret unit on the Art Bell show?" I asked him.

"Cross?" he said. "Irate? Angry? You bet."

"What was your motive for doing it?" I asked.

"I didn't have any motive." Ed shrugged. "I didn't have any motive at all."

We continued driving. We were on the beach road.

"I moved here for the peace and the beauty," said Ed. "But, yes, over the horizon there are some very, very nasty things coming. Things will get grim. Things will get ugly. This is a good place to be when that happens."

"What's going to happen?"

"We're all going to die!" said Ed. He laughed.

But then he said he meant it.

"In the next decade, humanity will see some of the most catastrophic changes to civilization it's seen in all of its recorded history. Earth changes. Biblically prophetic types of things."

"Like plagues?" I asked.

"No, that's minor," said Ed.

"Worse than plagues?"

"Diseases will ravage humankind, but I'm talking about actual Earth changes and I'm not kidding."

"Volcanoes and earthquakes?"

"The axis of the Earth will wobble and that'll shake up the oceans," said Ed. "Geophysically, we're in for Mr. Toad's Wild Ride within the next decade."

"These are things you've psychically viewed?" I asked.

"Many, many times," said Ed.

"Prudence says it was turning on the TV one day in Atlanta and seeing you that first got her interested in remote viewing," I said.

There was a silence. I wanted to gauge Ed's reaction to hearing the name Prudence. So many terrible things had happened, I was curious to see if he would flinch, but he didn't. Instead he turned vague.

"Most of the people practicing remote viewing on the streets today are either my students or students of my students," he said.

This was true. Although many of Ed's former military

associates eventually set up their own training schools after the unit's closure, Ed ran a campaign implying that many of the other secret psychics were psychically inferior to him. It worked. While Ed's house in Maui is in a fabulously opulent gated community near the beach, some of his former colleagues—like psychic Sergeant Lyn Buchanan—are compelled to struggle through as computer engineers, and so on. Lyn Buchanan is a legendary figure on the UFO circuit, but his gentle personality has denied him the opportunity to carve himself a niche in the increasingly cutthroat psychic spying private sector.

Prudence wanted Ed to teach her how to be a psychic spy, "But Ed didn't have any openings," she said. "He was booked solid for two years straight. Everybody wanted to be a psychic spy like Ed Dames."

So she settled for second best—an Atlanta-based lecturer in political science. His name was Dr. Courtney Brown.

Courtney Brown's credentials were impressive. He may not have been a top-level military spy, but he was an academic from a well-regarded university whose "vision statement," as outlined in their prospectus, was "to excel at discovery, generate wisdom, instill integrity and honor, set standards followed by others, be sought and prized for its opinions, and make discoveries that benefit the world."

"It was amazing to me," said Prudence. "Dr. Courtney Brown was just about Ed Dames's very first ever civilian student, then he set up his own training school, the Farsight Institute, in Atlanta. I was in Atlanta. I was living in the only city outside of L.A. where you could get training in remote viewing. So I signed up right away!"

Dr. Courtney Brown is handsome and clever, doe-eyed and tweedy. Having taken an eight-day one-on-one psychic-spying course with Ed Dames, he began teaching his version of the Dames method to scores of students.

He and Prudence became great friends. She ran his web site. Together they sat in Dr. Brown's basement and psychically spied on their favorite targets, aliens and mythical beasts and so on, the same fantastical things that Ed Dames used to remote view inside the military unit.

In July 1996, Prudence got a call from Art Bell. His millions of listeners had gone crazy for Ed Dames and were keen to hear anything related. Was Dr. Brown available to appear on his show?

"Every day was a new adventure," Prudence told me, "but this was the greatest adventure so far."

On the show, Art Bell asked Courtney Brown if he agreed with Major Dames about the "massive numbers of babies dying" and the imminent "tremendous winds on Earth."

COURTNEY BROWN: There definitely are climactic changes coming.

ART BELL: Like what?

COURTNEY BROWN: Within our children's lifetime we will start entering a *Mad Max* scenario. It's quite clear at this point that civilization has to hunker down and go into underground shelters.

ART BELL: *Underground shelters,* Professor Brown?

COURTNEY BROWN: Yes. The population comes apart. The political systems fall apart. There are roving gangs on the surface. The population basically survives in underground bunkers. And not everybody gets to go in the bunkers. Most people have to slug it out on the surface.

ART BELL: Well, excuse me if I say holy smoke, Dr. Brown. If you knew how much what you just said sounds like what Major Dames said, I guess you'd probably start digging.

The civilians who had trained under Ed Dames seemed to inherit their teacher's disdain for his former colleagues. On the Art Bell show, Courtney Brown said they weren't intellectually equipped to engage with the more profound by-products of their divinations. For instance, if the CIA asked a psychic spy to hunt for Saddam Hussein, and while psychically creeping through a Baghdad palace the spy chanced upon an extraterrestrial hiding in the shadows, he'd just keep on walking until he found the dictator. Surely, Courtney Brown suggested to Art Bell's listeners, any psychic spy worth his salt would stop and engage with the extraterrestrial, but oh no, not the military psychics. Art Bell agreed that this seemed crazy—and talk about wasted opportunities.

ART BELL: You did a serious professional project of Mars, didn't you?

COURTNEY BROWN: Well, I studied two ET species—a species called the Greys, and the Martians. Long ago, at the time when dinosaurs roamed on earth, there was an ancient Martian civilization. . . .

When the Martian civilization was wiped out by some planetary cataclysm on Mars back during dinosaur times, explained Dr. Courtney Brown, "the Galactic Federation sanctioned a rescue group of Greys" to save them.

"Many Martians were rescued," he said.

"Taken off planet?" asked Art Bell.

"Yes," said Courtney Brown. "But they are now in underground caverns back on Mars. They're happy to have been rescued, but they'd love to have been brought to Earth instead. The problem is, they're basically on a dead planet. They must leave. They're between a rock and a hard place. They've got to leave Mars. They must come here. But *this* planet is populated with an aggressive, hostile human species that has movies about invasions from Mars, and the Martians themselves are terrified. The remote viewing results on this are absolutely unequivocal."

Courtney Brown said that the Martians would certainly be arriving on Earth within two years. Art Bell immediately asked the question that was presumably haunting the more right-wing antiimmigration listeners among his audience:

ART BELL: Important question. How many Martians are there?

COURTNEY BROWN: It's not going to cause a population problem. We're probably talking enough to populate a reasonable city.

ART BELL: That's a small number, really.

COURTNEY BROWN: You may say, what is the incentive? Why should we help them? People have actually said to me, "Forget the altruism of us having a good name in the galaxy. Why should we help anybody? We had trouble accepting Cambodian and Vietnamese refugees at the end of the Vietnam War, so why should we help Martians of all people?"

Courtney's answer to these Earthling isolationists: forget altruism. The Martians have a "one-hundred-and-fifty-year technological advantage over us. Imagine if somebody like Saddam Hussein says to them, 'Hey! You want a place to land? Just come on over here.'"

This was why, Courtney Brown stated, with some urgency in his voice, it was imperative for the United States government to seize the opportunity and "get those Martian ships under NATO command. Get those Martians in through the proper immigration processes."

At this point, Art Bell expressed concern that "desperate people do desperate things." Even if the Martians were inherently peaceful, perhaps their hopeless living conditions inside the caves of Mars might render them unexpectedly and ungratefully violent when the Americans came to save them. Isn't that essentially what had happened in Grenada and Vietnam?

Courtney Brown assured him that he understood his concern, but that this would not happen.

Prudence thought Courtney did brilliantly on the Art Bell show.

"Courtney's charisma leaped off the air waves and into your lap," she said. "You could feel his sincerity and tenderness as he spoke his words."

And, as Prudence listened to the show that night, her telephone rang.

"Pru," said the voice. "It's Wolfie."

Wolfie was, according to Prudence, the Internet nickname of a woman named Suzy (not her real name), a friend of Houston-based radio newsreader Chuck Shramek. Prudence had met Suzy and Chuck in an Internet chat room; they had exchanged e-mails but never spoken.

"Pru," said Suzy, "there's something you have to see. Chuck got a picture of the Hale-Bopp comet, and there's something next to it. I'm going to send it to you."

At that moment Prudence's e-mail icon flashed. She opened the attachment to find a photograph. Chuck, Suzy said, had taken it from a telescope in his back garden. He was an amateur astronomer. Next to the Hale-Bopp comet, to the right of the frame, there seemed to be some kind of object.

When Prudence saw this photograph, she cried.

"The companion object," she told me, "glowed brighter than any star."

In the days that followed, Prudence and Courtney Brown and Courtney's other students began to work seriously on psychically viewing the Saturn-shaped object next to the Hale-Bopp comet.

"And we found," said Prudence, "that it was artificial. And it wasn't a mistake on Chuck's camera. It was an actual object. And it was alien in origin. It looked like a huge round

Chuck Shramek's photograph.

metal object and it had all these dents in it. Concave indentations. And it had antennae sticking out, tubes sticking out. And it was coming right toward us! And we were so excited. Courtney Brown phoned Art Bell right away."

On November 14, 1996, Art Bell announced two guests on his show: Chuck Shramek and Courtney Brown.

ART BELL: Chuck, welcome to the program.

CHUCK SHRAMEK: Thanks, Art, great to be here.

ART BELL: You are an amateur astronomer, right?

CHUCK SHRAMEK: Have been since I was eight years old. Now I'm forty-six.

ART BELL: So not such an amateur!

CHUCK SHRAMEK: Ha ha ha!

Chuck began to describe his photograph, how he had come to take it, how his heart had started pounding when he realized that the object—the "companion" to Hale-Bopp—wasn't a star, because he checked his star chart and there was no star like that in the comet's vicinity.

CHUCK SHRAMEK: This is a big thing. And there appear to be Saturn-like rings. This is amazing.

ART BELL: What could it be?

CHUCK SHRAMEK: Well, I think that might be an area for Courtney to get into. I have no idea.

ART BELL: So there you go, that's Chuck in Houston. We're going to ask Courtney Brown what it's all about. Maybe he can help. I suspect he can.

After the break, Courtney Brown offered the knockout—the result of his and Prudence's and the Farsight Institute's psychic study of Chuck Shramek's photograph.

ART BELL: I've seen the Hale-Bopp photograph and it really is odd. There's something really big out there. I've no idea what it is, but whatever it is, it's real. Well, Professor, what the hell is it?

COURTNEY BROWN: I'm willing to tell you. Do you want me to tell you?

ART BELL: Tell me.

Courtney attempted to sound scientific and levelheaded but he was unable to conceal his excitement.

COURTNEY BROWN: The information I'm about to give you is so far reaching, so incredible, you're going to be saying, How could this be? Remember, I'm a Ph.D.

ART BELL: Right.

COURTNEY BROWN: This object is approximately four times the size of the planet Earth, and it's headed our way. It apparently has tunnels in it. And it is moving by artificial means. It is under intelligent control. It's a vehicle. And there is a *message* coming from it.

ART BELL: Oh boy! There's a message coming from it?

COURTNEY BROWN: These beings are trying to communicate with us. This object is *sentient*. It is alive. It is *knowing*. It's something like the obelisk from *2001: A Space Odyssey*. It has hallways in it. This is good news. Our time of ignorance, our time of darkness, is coming to a close. We are entering a time of *greatness*. There are more of them coming!

ART BELL: *What?*

COURTNEY BROWN: My Lord . . . My Lord . . .

ART BELL: There are *more* of these coming? Folks, this is not a fake *War of the Worlds* broadcast. This is breaking news. I feel like I have been hit by a sledgehammer.

COURTNEY BROWN: Art, this one is real.

There was a short silence then, and then Art Bell spoke, and his voice trembled slightly.

ART BELL: Somehow I always felt I'd be on hand for this.

That night, Art Bell's web site crashed with the volume of traffic—listeners trying to log on so they could view Chuck Shramek's photograph. Finally, in just a few months' time—approximately mid March 1997, in fact—the Martians were coming.

An extraordinary thing about the Internet is how it can freeze moments in time. If you look hard enough, you can find the thoughts of some of Art Bell's listeners that night as they typed passionately with the radio playing in the background:

> Is this really happening? Oh man, this is incredible!! On Art Bell it was announced that some astronomers are now SUDDENLY seeing a Saturn-like huge object near our Comet Hale-Bopp! It is under intelligent control and it is connected to ETs!!

> Dear Friends,
>   As this incredible news is breaking I am typing wildly.
>   FLASH!!! Moving toward Earth a celestial object four times the size of Earth tracking right behind Comet Hale-Bopp; a ringed sphere, self-emitting light source, surface uniformly smooth and luminescent.
>   Is this the coming of the anti-Christ?

Prudence went on Art Bell too, a few days later, to clarify her psychic findings about the Hale-Bopp companion. She and Courtney were inundated with phone calls and e-mails.

"Thousands of e-mails," said Prudence. "We sent out a standard response to many of them because you just can't answer the whole world. You have to pick and choose."

One e-mail, among the thousands of others, seemed particularly odd to Prudence. It asked, "Will the companion raise us to the level above human?"

Prudence stared at this e-mail for a moment, then sent the standard reply: "Thank you for your interest in the Farsight Institute. Here's our upcoming class schedule. . . ."

In a pristine white house in a very rich suburb of San Diego, California, in mid-March 1997, a former music teacher from Texas named Marshall Applewhite turned on his video camera, pointed it at himself, and said, "We're so excited we don't know what to do because we're about to reenter the level above human!"

He turned the video camera away from himself to a room full of people. They were all dressed exactly the same, in buttoned-up uniforms of their own design, like something out of *Star Trek,* with a patch on the arm that read HEAVEN'S GATE AWAY TEAM.

They were all, like Marshall Applewhite, grinning.

"Heaven's Gate *Away* Team!" said Marshall Applewhite into the video camera. "That's exactly what that means to us. We've been away, and now we're going back. I'm very proud of these students of the evolutionary level above human. They're about to leave, and they're *excited* about leaving!"

Someone from this group had posted a message on their web site. It read: "Red Alert! Hale-Bopp brings closure to Heaven's Gate."

The web site also included a link to Art Bell's site.

Marshall Applewhite and his thirty-eight disciples went to a local restaurant for their last supper. They all ordered exactly the same thing from the menu—iced tea, salad with tomato vinaigrette dressing, turkey, and blueberry cheese-cake.

Then they returned to their communal home.

A few nights later, as Hale-Bopp drew close enough to Earth to be seen with the naked eye, Prudence stood on the balcony of a Holiday Inn in Atlanta and arched her neck uncomfortably to see over the trees, the iron railing digging into her chest. And then she saw the comet.

"It was so beautiful," she said.

"But it was by itself," I said.

"It was by itself," said Prudence. "I was just standing there, trying to figure out where the companion object went, and then someone came running up the stairs."

Thirty-nine people had died.

Marshall Applewhite and his thirty-eight disciples had all put on the exact same Nike sneakers. Each one put a roll of quarters in his pocket. They lay down on their bunk beds and each took a lethal cocktail of sedatives and alcohol and painkillers because they believed that doing so would get them a ride to the level above human on Prudence and Courtney's Hale-Bopp companion object.

"It was awful," said Prudence. "It was . . ."

She fell silent and put her head in her hands, staring off into the distance.

"They believed they were going to join the companion object to the comet," she said.

"Hmm," I said.

"All those people," she said.

"Uh," I said.

"It's kind of stressful to talk about," she said. "I don't really know what to say."

"I guess you weren't to know that all the excitement would, uh, lead to a mass suicide," I said.

"You'd think that if you're a remote viewer you should have been able to figure that out ahead of time," said Prudence.

Chuck Shramek—the man who took the "companion" photograph—died of cancer in 2000. He was forty-nine. After his death, a childhood friend of his named Greg Frost told *UFO Magazine* that Chuck had always been an inveterate prankster: "I was there on one occasion when he ran his voice through a filter that made him sound like Zontar the Warp Master while he communicated with some gullible ham-radio operators. Chuck had convinced a whole flock of them that he was a space alien from Venus."

My guess is that Chuck Shramek heard Ed Dames and then Courtney Brown on Art Bell and decided to play a trick on the remote viewers. So he doctored a photograph and got his friend to telephone Prudence. If this is what happened, I have no idea whether Suzy was in on the scam.

Art Bell banned Prudence and Courtney Brown from ever

again appearing on his show. Major Ed Dames is still a regular and popular guest. He is routinely introduced by Art Bell as "Major Edward A. Dames, U.S. army, retired now, a decorated military intelligence officer, an original member of the U.S. army prototype remote-viewing training program, the training and operations officer for the Defense Intelligence Agency, or DIA, psychic intelligence, or PSIINT collection unit. . . ."

Military acronyms are truly mesmerizing.

Ed's most recent appearance on Art Bell at the time of this writing was in the spring of 2004. He told the listeners, "Now this is important. Before everybody goes to bed, listen to this. When you see one of our space shuttles being forced to land because of a meteor shower, that is the beginning of the end. That is the *harbinger*. Immediately after that will begin some drastic geophysical changes in the Earth, resulting in a wobble and possibly an entire pole shift—"

"God!" interrupted Art Bell. "There will be some who live through this, Ed? Or will no one live through it?"

"We're looking at a couple of billion people who are going to get crisped," Ed replied.

I have noticed, however, a certain irreverence creeping into Art Bell's more recent interviews with Major Dames. Nowadays, amid the mesmerizing military acronyms, Art Bell sometimes refers to Major Dames as "Dr. Doom."

According to Prudence, Dr. Courtney Brown's Farsight Institute dwindled from thirty-six students to twenty students to eight students to no students at all during the months that followed the suicides (although it has since recovered). He stopped giving interviews. He hasn't spoken about what happened for seven years. (I think he went on Art Bell one more time to be shouted at.) I visited him in the spring of 2004.

He still lives in Atlanta. He is very thin now. He took me into his basement.

"Heaven's Gate?" he said.

He acted for a moment as if he couldn't quite remember who they were. He was wearing a tweed jacket with leather patches on the elbows.

"Heaven's Gate?" he said again. His look suggested that he had the memory of a vague academic and that I should bear with him for a moment.

"Oh!" he said. "Oh, yes. That was an interesting group. They were eunuchs. That's what I read in the newspaper. They castrated themselves and eventually they killed themselves."

Dr. Brown fell silent for a moment.

"It was like Jim Jones," he said. "Their leader was probably a crazy type of guy who was getting older and, seeing that his group was going to unravel in front of him, he was probably looking for some opportunity to finalize it."

Dr. Brown took off his glasses and rubbed his eyes.

"Eunuchs!" He chuckled dryly and shook his head. "That's pretty heavy psychological control to get people to castrate themselves, and eventually he had them all kill themselves as well, looking for an opportunity. You know, uh. That was an interesting group. That was a wild, wild group. That was a crazy group. That was a . . . that was a tragedy waiting to happen."

Dr. Brown made me some herbal tea.

He said, "You have to understand, I'm an academic. I'm

not trained in dealing with masses of people. I found out through the school of hard knocks that it is better not to deal with masses of people. It's not that they don't deserve the information but they really react in very strange ways. They get panicky and excited, or overexcited, and it is so easy for academics to forget that. We're trained in math. We're trained in science. We're not trained in the masses."

He paused.

"The public is extremely wild," he said, "uncontrollably wild."

Then he shrugged his shoulders.

"You have to understand," he said, "I'm an academic."

7. THE PURPLE DINOSAUR

If you walk about five hundred yards down the road from the Fort Bragg goats, you come to a large, modern, gray brick building with a sign at the front that reads C COMPANY 9TH PSYOPS BATTALION H-3743.

This is the army's Psychological Operations headquarters.

In May 2003, a little piece of the First Earth Battalion philosophy was put into practice, by PsyOps, behind a disused railway station in the tiny Iraqi town of al-Qā'im, on the Syrian border, shortly after President Bush had announced "the end of major hostilities."

The story begins with a meeting between two Americans—a *Newsweek* journalist named Adam Piore and a PsyOps sergeant named Mark Hadsell.

Adam was traveling in a PsyOps Humvee, driving into the town of al-Qā'im, past the coalition checkpoints, past the main road sign, which was shot up and dilapidated and now read A Q M. They pulled up in front of a police station. It was Adam's second day in Iraq. He knew virtually nothing about the country. He badly needed to urinate, but was worried that if he peed in front of the police station or in the

117

bushes, he might offend someone. What was the protocol regarding public urination in Iraq? Adam mentioned his concern to the PsyOps soldier sitting next to him in the Humvee. This was PsyOps' job—to understand and exploit the psyche and the customs of the enemy.

"Just go on the front tire," the soldier said to Adam.

So Adam jumped out of the Humvee, and that's when Psy-Ops sergeant Mark Hadsell wandered over and asked him if he wanted to be shot.

Adam was telling me this story two months later, back in the *Newsweek* offices in New York. We were upstairs in the boardroom, which was decorated with blowups of recent *Newsweek* covers: a masked Islamic fundamentalist with a gun under the headline WHY THEY HATE US, and President and Mrs. Bush in the White House garden under the headline WHERE WE GET OUR STRENGTH. Adam is twenty-nine, he looks younger, and he trembled a little as he recounted the incident.

"So that's how I met the guy," said Adam. He laughed. "He said did I want to get shot? So I quickly zipped up. . . ."

"Was he smiling as he said it?" I asked.

I pictured Sergeant Hadsell, whoever he was, with a big, friendly smile on his face asking Adam if he wanted to be shot.

"No," Adam said. "He just said, 'Do you want to be shot?'"

Adam and Sergeant Hadsell ended up friends. They bunked together in the PsyOps squadron command center in a disused train station in al-Qā'im, and borrowed DVDs from each other.

"He's a very gung-ho type of guy," said Adam. "The

squadron commander used to call him Psycho Six, because he was always ready to go in with firepower. Ha! He once told me that he pointed a gun at someone and pulled the trigger, and the gun wasn't loaded, and the guy peed in his pants. I don't know why he told me that story, because I didn't think it was funny. In fact I thought it was somewhat twisted and disturbing."

"Did *he* think it was funny?" I asked Adam.

"I think he thought it was funny," Adam said. "Yeah. He was an American-trained killer."

The people of al-Qā'im didn't know that Baghdad had fallen to coalition troops, so Sergeant Hadsell and his PsyOps unit were there to distribute leaflets bearing this news. Adam was tagging along, covering the "end of major hostilities" from the PsyOps' perspective.

May 2003 was a pretty calm month in al-Qā'im. By the end of the year, U.S. forces would be under frequent guerrilla bombardment in the town. In November 2003, one of Saddam Hussein's air defense commanders—Major General Abed Hamed Mowhoush—would die under interrogation right there at the disused train station. ("Natural causes," said the official U.S. military statement. "Mowhoush's head was not hooded during questioning.")

But for now it was peaceful.

"At one point," Adam said, "somebody ran by and grabbed a pile of leaflets. Hadsell talked about how important it was, the next time that happened, to find the guy and punish him so he wouldn't do it again. That was probably to do with studying Arabic culture. You have to show that you're strong."

One night, Adam was hanging out in the squadron command center when Sergeant Hadsell wandered over to him. Hadsell winked conspiratorially and said, "Go look out by where the prisoners are."

Adam knew that the prisoners were housed in a yard behind the train station. The army had parked a convoy of shipping containers back there, and as Adam wandered toward them he could see a bright flashing light. He could hear music too. It was Metallica's "Enter Sandman."

From a distance it looked as though some weird and slightly sinister disco was taking place amid the shipping containers. The music sounded especially tinny, and the light was being joylessly flashed on and off, on and off.

Adam walked toward the light. It was really bright. It was being held by a young American soldier, and he was just flashing it on and off, on and off, into the shipping container. "Enter Sandman" was reverberating inside the container, echoing violently around the steel walls. Adam stood there for a moment and watched.

The song ended and then, immediately, it began again.

The young soldier holding the light glanced over at Adam. He continued flashing it and said, "You need to go away now."

"Ha!" said Adam to me, back in the *Newsweek* offices. "That's the term he used. 'You need to go away.'"

"Did you look inside the container?" I asked him.

"No," said Adam. "When the guy told me that I had to go away, I went away." He paused. "But it was kind of obvious what was going on in there."

Adam called *Newsweek* from his cell phone and pitched

them a number of stories. Their favorite was the Metallica one.

"I was told to write it as a humorous thing," said Adam. "They wanted a complete playlist."

So Adam asked around. It turned out that the songs being blasted at prisoners inside the shipping container included Metallica's "Enter Sandman"; the soundtrack to the movie *XXX*; a song that went "Burn Motherfucker, Burn"; and, rather more surprisingly, the "I Love You" song from *Barney & Friends,* the Barney the Purple Dinosaur show, along with songs from *Sesame Street.*

Adam e-mailed the article to New York, where a *Newsweek* editor phoned the Barney people for a comment. He was put on hold. The on-hold music was the Barney "I Love You" song.

The last line of the article, written by the *Newsweek* editor, was: "It broke us too!"

I first learned about the Barney torture story on May 19, 2003, when it ran as a funny, "and finally . . ." type of item on NBC's *Today* show:

ANN CURRY (news anchor): U.S. forces in Iraq are using what some are calling a cruel and unusual tool to break the resistance of Iraqi POWs, and trust me, a lot of parents would agree! Some prisoners are being forced to listen to Barney the Purple Dinosaur sing the "I Love You" song for twenty-four straight hours. . . .

NBC cut to a clip from *Barney,* in which the purple dinosaur flopped around amid his gang of ever-smiling stage-school kids. Everyone in the studio laughed. Ann Curry put on a funny "poor little prisoners" kind of voice to report the story.

ANN CURRY: . . . according to *Newsweek* magazine. One U.S. operative told *Newsweek* that he listened to Barney for forty-five minutes straight and never wanted to go through *that* again!

STUDIO: (laughter)

Ann Curry turned to Katie Couric, her cohost.

ANN CURRY: Katie! Sing it with me!

KATIE COURIC (laughing): No! I think after about an hour they're probably spilling the beans, don't you think? Let's go outside to Al for the weather.

AL ROKER (weatherman): And if Barney doesn't get 'em, they switch to the Teletubbies, and that crushes 'em like a *bug . . .* !

It's the First Earth Battalion! I thought.

I had no doubt that the notion of using music as a form of mental torture had been popularized and perfected within the military as a result of Jim Channon's manual. Before Jim came along, military music was confined to the marching-band type of arena. It was all about pageantry and energizing the troops. In Vietnam, soldiers blasted themselves with Wagner's "Ride of the Valkyries" to put themselves in the mood for battle. But it was Jim who came up with the idea of loudspeakers being used in the battlefield to broadcast "discordant sounds" such as "acid-rock music out of sync" to confuse the enemy, and the use of similar sounds in the interrogation arena.

Jim got these ideas in part, as far as I could tell, after he met Steven Halpern, the composer of ambient CDs such as *Music for Inner Peace,* in 1978. And so I called Jim right away.

"Jim!" I said. "Would you say that blasting Iraqi prisoners with the theme tune to *Barney* is a legacy of the First Earth Battalion?"

"I'm sorry?" said Jim.

"They're rounding people up in Iraq, taking them to a shipping container, and blasting them repeatedly with children's music while repeatedly flashing a bright light at them," I said. "Is that one of your legacies?"

"Yes!" Jim said. He sounded thrilled. "I'm so pleased to hear that!"

"Why?" I asked.

"They're obviously trying to lighten the environment," he said, "and give these people some comfort, instead of beating them to death!" He sighed. "Children's music! That will make the prisoners more ready to divulge where their forces are and shorten the war! Damn good!"

I think Jim was imagining something more like a day care center than a steel container at the back of a disused railway station.

"I guess if they play them *Barney* and *Sesame Street* once or twice," I said, "that's lightening and comforting, but if they play it, say, fifty thousand times into a steel box in the desert heat, that's more . . . uh . . . torturous?"

"I'm no psychologist," said Jim, a little sharply.

He seemed to want to change the subject, as if he was in denial about the way in which his vision was being interpreted behind the railway station in al-Qā' im. He reminded me of a grandparent who wouldn't countenance the idea that his grandchildren would ever misbehave.

"But the use of music . . ." I said.

"That's what the First Earth Battalion *did,*" said Jim. "It opened the military mind to how to use music."

"So," I said, "it's all about getting people to talk in a . . . in a what?"

"A psychospiritual dimension," said Jim. "Besides the basic fear of being hit, we have a mental, spiritual, and psychic component. So why not use that? Why not go straight for the place where the *being* actually decides whether to say something or not?"

"So are you certain," I asked Jim, "given what you know about how your First Earth Battalion has disseminated its way into the fabric of the military, that blasting Iraqis with *Barney* and *Sesame Street* is one of your legacies?"

Jim thought about this for a moment and then he said, "Yes."

Christopher Cerf has been composing songs for *Sesame Street* for twenty-five years. His large Manhattan townhouse is full of *Sesame Street* memorabilia—photographs of Christopher with his arm around Big Bird, and so on.

"Well, it's certainly not what I expected when I wrote them," Christopher said. "I have to admit, my first reaction was, 'Oh my gosh, is my music really that terrible?'"

I laughed.

"I once wrote a song for Bert and Ernie called 'Put Down the Ducky,'" he said, "which might be useful for interrogating members of the Ba'ath Party."

"That's very good," I said.

"This interview," Christopher said, "has been brought to you by the letters *W*, *M*, and *D*."

"That's very good," I said.

We both laughed.

I paused.

"And do you think that the Iraqi prisoners, as well as giving away vital information, are learning new letters and numbers?" I said.

"Well, wouldn't that be an incredible double win?" said Christopher.

Christopher took me upstairs to his studio to play me one of his *Sesame Street* compositions, called "Ya! Ya! Das Is a Mountain!"

"The way we do *Sesame Street*," he explained, "is that we have educational researchers who test whether these songs are working, whether the kids are learning. And one year they asked me to write a song to explain what a mountain is, and I wrote a silly yodeling song about what a mountain was."

Christopher sang me a little of the song:

> Oompah-pah!
> Oompah-pah!

Ya! Ya! Das is a mountain!
Part of zee ground zat sticks way up high!

"Anyway," he said, "forty percent of the kids had known what a mountain was *before* they heard the song, and *after* they heard the song, only about twenty-six percent knew what a mountain was. That's all they needed. You don't know what a mountain is now, right? It's gone! So I figure if I have the power to suck information out of people's brains by writing these songs, maybe that's something that could be useful to the CIA for brainwashing techniques."

Just then, Christopher's phone rang. It was a lawyer from his music publisher, BMI. I listened in to Christopher's side of the conversation.

"Oh really?" he said. "I see . . . Well, theoretically they have to log that and I should be getting a few cents for every prisoner, right? Okay. Bye-bye."

"What was that about?" I asked Christopher.

"Whether I'm owed some money for the performance royalties," he explained. "Why not? It's the American thing to do. If I can write songs that drive people crazy sooner and more effectively than others, why shouldn't I profit from that?"

This was why, later that day, Christopher asked Danny Epstein—who has been the music supervisor of *Sesame Street* since the very first program was broadcast, in July 1969—to come to his house. It would be Danny's responsibility to collect the royalties from the military if they proved negligent in filing a music-cue sheet.

For an hour or so, Danny and Christopher attempted to

calculate exactly how much money Christopher might be due if—as he estimated—his songs were being played on a continuous loop in a shipping container for up to three days at a time.

"That's fourteen thousand times or more over three days," said Christopher. "If it was radio play I'd get three or four cents every time that loop went through, right?"

"It would be a money machine," concurred Danny.

"That's what I'm thinking," said Christopher. "We could be helping our country and cleaning up at the same time."

"I don't think there's enough money in the pool to pay for that rate," said Danny. "If I'm going to negotiate for ASCAP [American Society of Composers, Authors and Publishers], I'd say it would come in the category of a theme or jingle rate, some kind of knockdown. . . ."

"Which is an appropriate term because there's evidence that the prisoners are being knocked down as they listen to the music," said Christopher.

We all laughed.

The conversation seemed to be shifting uneasily between satire and a genuine desire to make some money.

"And that's just in one interrogation room," said Danny. "If there's a dozen rooms, you're talking money. This is non-sponsored?"

"That's a good question," said Christopher. "It's state sponsored, I think. Would I get more money if it is or it isn't?"

"Now, would we have a special rate for Mossad?" said Danny.

We laughed.

"I think we ought to collect royalties," said Christopher. "If I'd written the songs directly for the army, they would pay me, right?"

"No," said Danny. "You'd be a work for hire. You'd be employed by them."

"Well, I'm not a work for hire in *this* case," said Christopher.

"I'm not so sure," said Danny. "As a citizen, you have to work for hire, if the military needs you."

"Well, they could have asked me to volunteer," said Christopher.

He was more serious now. Danny took off his glasses and rubbed his eyes.

"Wanting money for the use of your music in a time of crisis," he said, after a moment, "seems a little shabby to me."

And the two men collapsed in helpless laughter.

In the late autumn of 2003, after many faxes and e-mails had gone back and forth, and I had been security screened by various offices within the Pentagon and the American embassy, PsyOps consented to show me their CD collection.

Adam Piore, the *Newsweek* journalist, had said that the list of songs blasted at the prisoners had been chosen here at PsyOps headquarters. The collection was housed in a series of radio-production suites inside a low brick building in the middle of Fort Bragg, some five hundred yards up the road from where, it was rumored, Goat Lab was situated. I kept looking out windows in the hope of spotting dazed or hobbling goats, but there was none in evidence.

PsyOps began by showing me their sound-effects CDs.

"Primarily deception," explained the sergeant who guided me during this portion of the day, "designed to make enemy forces think they're hearing something that doesn't exist."

One sound-effect CD was labeled "Crazy Woman Says 'My Husband's Never Liked You.'"

"We purchased a job lot," explained the sergeant.

We laughed.

"Many Horses Galloping By" read another, and we laughed again and said this would have been deployable three hundred years ago, but not now.

Then he played me an applicable one: "Tank Noises."

The radio suite filled with the rumblings of tanks. They seemed to be coming from everywhere at once. The sergeant explained that sometimes PsyOps hide behind a hill to the east of the enemy and blast their tank noises as the real tanks rumble in, more quietly, from the west.

Then he showed me their Arabic music CDs ("Our analysts and our specialists are familiar with what may be popular and culturally relevant, and we purchase that music in order to appeal to the population"), followed by their collection of Avril Lavigne and Norah Jones CDs.

"How might Avril Lavigne be deployed in hostile countries?" I asked.

There was a silence.

"In some parts of the world Western music is popular," he replied. "We try to stay current."

"Who chooses the playlist?" I asked.

"Our analysts," he said, "in conjunction with our specialists."

"Which countries?" I asked.

"I don't want to go into that," he said.

My tour of PsyOps was a well-rehearsed whirlwind—the same tour as a visiting dignitary or a congressman would get. A PsyOps soldier knows how to design a leaflet and burn a CD and operate a loudspeaker and take a photograph and snap into formation for the official tour.

They showed me their radio studios and their TV studios and their archive library, with shelves full of videos labeled "Guantanamo Bay," and so on. I noticed a poster on a wall reminding the soldiers of PsyOps of their official functions: "Surrender appeals. Crowd control. Tactical deception. Harassment. Unconventional warfare. Foreign internal defense."

They showed me their leaflet-printing presses, and their canisters. These are dropped from planes and are designed to split open in midair, and then tens of thousands of leaflets float down into enemy territory.

The Americans have always been better than the Iraqis at the leaflets. Early on in the first Gulf War, Iraqi PsyOps dropped a batch of their own leaflets on U.S. troops, designed to be psychologically devastating. They read, "Your wives are back at home having sex with Bart Simpson and Burt Reynolds."

Then I was led into a PsyOps conference room where I was introduced to the specialists and the analysts. Some were in uniform. Others looked like friendly eggheads, bespectacled and in business suits.

The specialists showed me some of their leaflets that had floated down from PsyOps helicopters into Iraqi forces just a month or two earlier. One read, "Nobody benefits from the use of Weapons of Mass Destruction. Any unit that chooses to use Weapons of Mass Destruction will face swift and severe retribution by coalition forces."

"This product," explained a specialist, "is making a clear link between *their* unfulfilled need and our desired behavior."

"What do you mean?" I asked him.

"Their unfulfilled need," he said, "was that they didn't want to face severe retribution. And our desired behavior was that we didn't want them to use weapons of mass destruction."

I nodded.

"Our most effective products are the ones which link an unfulfilled need on their part with a desired behavior on our part," he said.

There was a silence.

"And weapons of mass destruction were *not* used on American forces," the specialist added, "so this leaflet may very well have been effective."

"Do you really . . . ?" I started. "Oh, nothing," I said.

I picked up another leaflet. It read, "You people aren't being fed. Your children are going hungry. While you live in squalor, Saddam's generals are so overweight and fat he has to fine them to keep them in fighting shape."

As I read this leaflet, I had a short conversation with a Psy-Ops analyst named Dave. He wasn't in uniform. He was a friendly, middle-aged man. What he said to me didn't seem particularly significant at the time, so I just nodded and smiled, and then I was bustled out of the conference room and into an oak-lined office where a tall, handsome man wearing khaki shook my hand and said, "Hi, I'm Colonel Jack N. . . ."

He blushed, disarmingly.

"N!" He laughed. "Middle name! Jack N. Summe. I'm the commander of the Fourth Psychological Operations Airborne, Fort Bragg, North Carolina."

"Are you in charge of all of PsyOps?" I asked him.

My hand was still being vigorously shaken.

"I'm in charge of the active-duty Psychological Operations group in the United States Army," he said. "Our job is to convince our adversaries to support U.S. policies and to make the battlefield a less dangerous place using multimedia techniques."

"Colonel Summe," I said. "What can you tell me about the deployment of *Barney* and *Sesame Street,* by PsyOps, inside shipping containers in al-Qā'im?"

Colonel Summe didn't miss a beat.

"I was at the Joint Staff headquarters and I took command of the Fourth PsyOps group on seventeen July, so I have not had the ability to operationally deploy into Iraq and find out at what level we're doing things." He paused to take a very short breath and continued, "We serve as force providers. When there is a requirement—a crisis requirement—we are tasked to send PsyOpers forward to support. When Psychological Operations deploys . . ."

Colonel Summe's words, delivered like machine-gun fire, swam around my head. I smiled and nodded blankly at him.

" . . . We are always in support of the commander. The senior commander or maneuver commander or area commander is never a PsyOps officer. We are always a support force. So when we attach that PsyOps force to a commander he may identify a use of Psychological Operations loudspeaker capability for that very reason. . . ."

I continued nodding. It was almost as if Colonel Summe wanted to tell me something, but he wanted to say it in such a way that he didn't want me to understand it. Maybe, I thought, as my mind drifted, and I glanced out the window to the lawn outside his office in the vain hope of spotting injured goats, he was performing some kind of PsyOp on me.

" . . . If we have combat forces in the field I would rather see our PsyOps capability being used to support those combat forces as opposed to some other mission like you've kind of outlined."

Then Colonel Summe coughed and he shook my hand again and he thanked me for my interest and I was bustled out the door.

8. THE PREDATOR

Martial arts master Pete Brusso, who teaches hand-to-hand combat at the Camp Pendleton Marine base near San Diego, has read Jim Channon's *First Earth Battalion Operations Manual* cover to cover. Just a week before I met Pete, in March 2004, he had had a long telephone conversation with Jim, during which they discussed how First Earth principles might be deployed in Iraq today. Pete had a "number of my operatives" in Iraq "right now," he told me.

We were cruising through Camp Pendleton inside Pete's $167,000 Hummer H1. His license plate read MY OTHER CAR IS A TANK. Pete's Hummer is something like a nightmare version of the *Chitty Chitty Bang Bang* car, in that it can swim, can effortlessly navigate the most treacherous terrain on the planet, and has a number of places scattered around the bodywork on which to mount one's weaponry. Pete turned the music up loud to demonstrate his state-of-the-art sound system. He played me a very loud, crystal clear, but weird song, which basically went *bling blong bling blong*.

"I COMPOSED THIS MYSELF," shouted Pete.

"WHAT?" I said.

Pete turned the music down.

"I composed this music myself," he said.

"It's interesting," I said.

"I'll tell you *why* it's interesting," said Pete. "It thwarts bugs. Someone's bugging this Hummer? Just turn up the music. The bugging device can't cope. Usually spies can take a bugging tape, strip out the music, and hear the conversation. Not with *this* music."

Pete does for the Marines at Camp Pendleton what Guy Savelli used to do for Special Forces at Fort Bragg. He teaches them martial arts techniques with a First Earth Battalion dimension. But, unlike Guy, Pete is a military veteran. He fought in Cambodia for ten months. His combat experience has made him sniffy about Guy's goat-staring capabilities. Violent goats don't come running at you on the battlefield. Guy's goat staring may be fabled, but it is basically a party trick.

Then Pete turned the music up loud and told me a secret, which I couldn't hear a word of, so he turned the music back down and told me it again. The secret was that he and Guy Savelli were now competitors. Military commanders have been considering a mandatory post-9/11 martial arts training program. The two sensei—Pete and Guy—were vying with each other for the contract. Pete basically said there was no contest. Would the military really want a *civilian*, like Guy, with his party tricks?

In short, Pete is a pragmatist. He's an admirer of the First Earth Battalion but has taken it upon himself to adapt Jim's ideas into practical applications for the battlefield Marine.

I asked Pete to give me an example of a practical application.

"Okay," he said. "There's a gang of insurgents standing in front of you. You are alone. You want to dissuade them from attacking you. What do you do?"

I told Pete I didn't know.

Pete said that the answer lay in the psychic realm—specifically the use of visual aesthetics to instill psychically in the enemy a disincentive to attack.

"Can you be more explicit?" I asked.

"Okay," said Pete. "What you do is grab one of them, rip out his eyeballs, and stab him in the neck, the blood squirts out like a fountain—really, a *fountain*—get the blood to spurt over his friends. Just punish the bejesus out of him, right there in front of his friends."

"Okay," I said.

"Or go for the lungs," said Pete. "Create a gaping chest wound. What you'll have then is lots of sucking in air and frothing. Or scrape a knife across the face. Here's a clever thing: Get your knife inside the clavicle. That's the collarbone. Once you're in there you can scrape most of the tissue away from that side of the neck. Separate his brain stem from the back of his neck. Doesn't take much movement, physics-wise." Pete paused. "What I'm doing, you see, is creating a powerful visual psychic disincentive for the other insurgents to attack me."

Pete turned the music up loud.

"THAT'S A . . ." I shouted.

"WHAT?" shouted Pete.

" . . . BROAD INTERPRETATION OF JIM'S IDEALS," I yelled.

Pete turned down the music again and shrugged as if to say, *There you go. That's warfare.*

We pulled up outside a hangar. A half-dozen of Pete's trainees were waiting for him. We wandered inside. Then Pete said, "Choke me."

"I'm sorry?" I said.

"Choke me," said Pete. "I'm old and fat. What can I do? Choke me. Right here."

Pete pointed at his neck.

"Now choke," he said, softly. "Choke. Choke."

"You know," I said to Pete, "I feel that neither of us has anything to prove here."

"Choke me," said Pete. "'Attack me.'"

When he said the words *attack me,* he did that quotation-mark thing in the air with his fingers, which angered me a little because it implied that I was incapable of mounting anything more than a figurative attack. I was indeed incapable, but I had known Pete for only a few minutes and I felt he was jumping to conclusions about me.

"If I do choose to choke you," I said, "what do you intend to do?"

"I'm going to interrupt your thought pattern," said Pete. "It'll take your brain three-tenths of a second to realize what is happening to you. And after that three-tenths of a second you'll be mine. I'm going to touch you and that's it. I'm not going to move even my feet. But I'm going to project myself into you, and you will fly."

"Well," I said. "If I do decide to choke you, will you bear in mind that I am not a Marine?"

"Choke me," said Pete. "Choke."

I looked behind me and I saw a number of sharp edges.

"*Not* into sharp edges," I said. "Not sharp edges."

"Okay," said Pete. "No sharp edges."

I raised my hands in readiness to choke Pete, and I was surprised to see how violently they were trembling. I had presumed until that moment that we were essentially kidding around, but the sight of my hands made me realize we weren't. At that moment of realization, the rest of my body caught up with my hands. I felt incredibly weak. I put my hands back down.

"Choke me," said Pete.

"Before I choke you," I said, "I would like to ask you one or two other questions."

"Choke me," said Pete. "Come on. Choke me. Just choke me."

I sighed, placed my hands around Pete's neck, and began to squeeze.

I didn't see Pete's hands move. All I know is that both my armpits, my neck, and my chest began to hurt enormously, all at once, and then I was flying, flying across the room, flying toward two Marines, who stepped gently out of the way, and then I was skidding, skidding like a sore ice-skater toward sharp edges, and I came to rest a few inches from those edges. I was in great pain but also impressed. Pete was truly a maestro of violence.

"Fuck," I said.

"Does it hurt?" asked Pete.

"Yes," I said.

"I know it does," said Pete. He seemed pleased. "It fucking hurts like hell, doesn't it?"

"Yes," I said.

"You felt fear," said Pete, "didn't you? Beforehand."

"Yes," I said. "I was debilitated with fear beforehand."

"Would you say that level of fear was abnormal for you?" asked Pete.

I thought about this.

"Yes and no," I said.

"Explain further," said Pete.

"I do sometimes experience fear when something bad is happening to me, or is about to happen to me," I explained, "but on the other hand, the *amount* of fear I felt in the run-up to the choking seemed unusual. I was definitely more scared than I ought to have been."

"You know why?" said Pete. "It wasn't you. It was me. It was thought projection. I was inside your head."

I was, he explained, a real-life plaything of a practical application of Jim Channon's vision. I was the Iraqi insurgent being sprayed with the fountain of blood emanating from the neck of his friend. I was the hamster. I was the goat.

And then Pete produced, from his pocket, a small yellow blob of plastic. It had pointed edges and smooth edges and a hole in the middle. It looked like a children's toy, albeit one with no obvious means of being fun. This yellow blob, Pete said, was his own design, but it was an embodiment of Jim Channon's vision, and it was being carried right now in the pockets of the Eighty-second Airborne in Iraq, and soon,

Pentagon willing, it would be in the pocket of every soldier in the United States Army. His blob, Pete said, "is friendly to the Earth, it has a spirit to it, it is as humane as you want it to be, the pointed bits go into people, it can snuff your life out in a heartbeat, and it looks a little bit funny. It is," he said, "the First Earth Battalion."

"What's it called?" I asked.

"The Predator," said Pete.

For the next hour or two, Pete hurt my chakra, in many, many ways, with his Predator. He grabbed my finger, placed it in the hole, and twisted it 180 degrees.

"You're mine now," he said.

"Stop hurting me," I said.

He grabbed my head, and stuck the pointed bit in my ear.

"For Christ's sake," I said. "Stop."

"This is a great Iraqi story, by the way," said Pete.

"The car thing?" I asked, getting up from the floor and brushing myself down.

"Yeah," said Pete.

"What does the Iraqi story have to do with putting the Predator in someone's ear?" I asked.

Pete began to tell me, but a Marine commander standing near us shook his head, barely perceptibly, and Pete fell silent.

"Sufficient to say," Pete said, "the Iraqi who didn't want to stand up stood up." He paused. "You want a bit of pain compliance?" he asked me.

"No," I said.

Pete rapidly rubbed the serrated edge of the Predator against a part of my temple, and, as I let out a bloodcurdling scream, he grabbed my fingers and squeezed them agonizingly against the smooth edge.

"STOP!" I yelled.

"Picture this scenario," said Pete. "We're in a bar in Baghdad and I want you to come with me. Are you coming now?"

"Stop hurting me all the time," I said.

Pete stopped and looked at his Predator fondly.

"What's cool about it," he said, "is that if you found it on the ground no one would know what it was, yet it has such lethality."

Pete paused. "Eyeballs," he said.

"NO!" I said.

"You can take eyeballs right out," said Pete, "with this bit."

On the thirty-fourth floor of the Empire State Building, in New York City, Kenneth Roth, the director of Human Rights Watch, realized that he was in an awkward situation. Ever since the Barney story had broken, journalists had been calling him for his comment. It was an engagingly surreal joke, but there was also a comforting familiarity to it. It was the comedy of recognition. If Barney was involved, the torture didn't sound *that* bad. In fact, an article in *The Guardian,* published on May 21, 2003, a newspaper that usually found little to be funny or upbeat about regarding the war in Iraq, said as much:

> What the former Fedayeen and Republican Guard are going through now is nothing. So they're being played the Barney song. At what time? Middle of the day? Meaning-

less. Only when you've been dragged from sleep before dawn, day after day for months on end, to enter Barney's Day-Glo world . . . only then do you know the full horror of the psychological warfare that is life with a pre-school child.

It had become the funniest joke of the war. Within hours of Adam Piore's *Newsweek* article appearing, the Internet was aflame with Barney torture-related wisecracks such as, "An endless loop of the theme song from *Titanic* by Celine Dion would be infinitely worse! They'd confess everything within ten minutes!"

And, from a different discussion group: "I think twelve hours of Celine Dion would be needed on the really tough ones!"

A third discussion group I saw had the following message posted on it: "Why didn't they just go all out and play them some Celine Dion? Now that would be cruel and unusual punishment!"

And so on.

Celine Dion's theme from *Titanic* was, in fact, being played in Iraq, albeit in a different context. One of PsyOps' first jobs, once Baghdad had fallen, was to seize Saddam-controlled radio stations and broadcast a new message— that America was not the Great Satan. One way in which they hoped to achieve this was by playing "My Heart Will Go On," over and over. How could a country that pro-duced melodies such as this be all bad? This sounded to me a lot like Jim Channon's vision of "sparkly eyes" and "baby lambs."

Adam Piore himself had told me that he was finding the impact of his Barney story quite baffling.

"It has had a tremendous amount of attention," he said. "When I was in Iraq my girlfriend called to tell me she'd seen it scroll across the CNN ticker. I didn't believe her. I thought there must be some mistake. But then Fox News wanted to interview me. Then I heard it was on the *Today* show. Then I saw it in *Stars & Stripes*."

"How did they report it?" I asked him.

"As humorous," said Adam. "Always as humorous. It was sort of outrageous to be in this shit hole up on the border in an abandoned train station, totally uncomfortable, unable to take showers, sleeping on cots, and when we finally got cable a couple of days later, scrolling across the screen is this . . . *Barney* story."

Kenneth Roth, of Human Rights Watch, could read the mood. He realized that if his responses to the journalists were overly austere, it would seem that he wasn't *getting* it. He would sound like a sourpuss.

So he said to journalists, myself included, "I have small kids. I can understand being driven crazy by the Barney theme song! If I had to have 'I Love You, You Love Me' played at high decibel over and over for hours, I might be willing to confess to anything as well!"

And the journalists laughed, but he would quickly add, "And I wonder what else is going on in those shipping containers while the music is being played! Perhaps the prisoners are being kicked around. Perhaps they're naked with a bag on their head. Perhaps they're chained and hanging upside down. . . ."

But the journalists rarely, if ever, included those possibilities in their stories.

By the time I met Kenneth Roth he was clearly sick of talking about Barney.

"They have," said Kenneth, "been very savvy in that respect."

"Savvy?" I said.

He seemed to be implying that the Barney story had been deliberately disseminated *just so* all the human-rights violations being committed in postwar Iraq could be reduced to this one joke.

I put this to him and he shrugged. He didn't know what was going on. That, he said, was the problem.

What I did know was that Sergeant Mark Hadsell, the PsyOps soldier who approached Adam Piore that night and said to him, "Go look out by where the prisoners are," had been given nothing more than a light reprimand for his indiscretion. Was Kenneth Roth right? Had Barney been chosen to torture people in Iraq simply because the dinosaur provided that powerful thing: a funny story for the people back home?

There is a room in a police building on top of a hill in Los Angeles that houses an array of pepper sprays and stun guns and malodorants—tiny capsules of powdered "fecal matter, dead mammals, sulfur, and garlic," which are "great at crowd dispersement" and "will gag a maggot." The man who showed me these things was Commander Sid Heal of the Los Angeles County Sheriff's Department. After the First

Earth Battalion's Colonel John Alexander, Sid is America's leading advocate of nonlethal technologies.

Sid and Colonel Alexander—"my mentor" Sid called him—frequently get together at Sid's house to test out various new electronic zappers on each other. If the two men are impressed, Sid introduces them into the L.A. law-enforcement arsenal. Then, like the now widespread Taser stun gun, the weapons sometimes spread throughout the entire U.S. police community. One day, somebody might calculate how many people are alive today, having not been shot dead by police officers, because of Sid Heal and Colonel Alexander.

Sid Heal had dedicated his life to researching new non-lethal technologies, so I assumed that he would know all about the Barney torture, but when I described to him what I knew—the flashing light, the repetitive music, the shipping container—he looked perplexed.

"I don't know why they're doing it," I said.

"I don't either," he said.

There was a silence.

"Do you think *they* know why they're doing it?" I asked.

"Oh yeah." Sid smiled. "I don't think anybody would go to those lengths to set up that elaborate a system without some ultimate scheme in mind. We don't experiment on each other. Not in our culture."

Sid fell silent. He thought about the Barney technique, and the accompanying flashing lights, and a startled look suddenly crossed his face.

"I suppose it could . . ." He paused and then said, "Nah."

"What?" I said.

"It could be the Bucha Effect," he said.

"The Bucha Effect?" I said.

Sid told me about the first time he heard about the Bucha Effect. It was in Somalia, he said, during the partially disastrous deployment of Colonel Alexander's Sticky Foam. The nonlethal-technology experts who had accompanied the foam to Mogadishu were understandably down in the dumps that night, and the talk got around to what might be the holy grail of these exotic technologies. It was then that a Lieutenant Robert Ireland spoke of the Bucha Effect.

It all began in the 1950s, Sid told me, when helicopters started falling out of the sky, just crashing for no apparent reason, and the pilots who survived couldn't explain it. They had just been flying around as normal and then suddenly they felt nauseated and dizzy and debilitated and they lost control of their helicopters and they went down.

So a Dr. Bucha was called in to solve the mystery.

"What Dr. Bucha found," said Sid, "was that the rotor blades were strobing the sunlight and when it reached the approximation of human brain-wave frequency it was interfering with the brain's ability to send correct information to the rest of the body."

As a result of Dr. Bucha's findings, new safety measures were introduced, such as tinted glass and helmet visors and so on.

"Believe me," said Sid Heal, "there are easier ways of doing sleep deprivation than going to all those great lengths. *Barney* music? Flashing lights? Sleep deprivation may be a part of it, but it's got to have some deeper hidden effect. My guess is that this is the Bucha Effect. My guess is that they're going for the amygdala.

"Picture this," said Sid. "You're walking down a dark hallway, and a figure jumps out in front of you, and you scream and jump back and all of a sudden you realize it's your wife. That's not two pieces of information," he said. "It's the same information being processed simultaneously by two different parts of the brain. The part where the judgment is takes three or four seconds. But the part that's *reactionary*—the amygdala—just takes a split second."

The quest for seizing that amygdala moment, those crushing seconds of unbearable, incapacitating shock, seizing those moments and not letting them go, dragging them out for as long as is operationally necessary, that, said Sid, is the aim of the Bucha Effect.

"It would be the ultimate nonlethal," he said.

"So," I said, "the Barney strobe-light musical torture inside a shipping container at the back of a railway station in al-Qā'im may in fact be the *ultimate* nonlethal?"

"I don't know anyone that's succeeded," said Sid. "The problem is that the threshold between it being effective and permanently disabling is so narrow I . . ."

Then Sid fell silent, I think because he realized that if he completed his sentence it would take his mind to a place he didn't want to go, a place where soldiers in Iraq didn't actually care, like he cared, about that threshold.

"But they might have succeeded," I said.

"They might have succeeded," said Sid, wistfully. "Yeah." Then he added, "But any sort of nonlethal weapon that would force compliance in interrogation wouldn't be appealing to us at all, because the resulting evidence couldn't be used in court."

"But they don't have those constraints inside a shipping container in al-Qā'im," I said.

"No, they don't," said Sid.

"Huh," I said.

"You know what you've stumbled into here?" said Sid.

"What?" I asked.

"The dark side," he said.

I left Sid and returned to the United Kingdom to find that I had been sent seven photographs. They were taken by a *Newsweek* photographer, Patrick Andrada, in May 2003, and were captioned, "An escaped detainee is returned to a holding area in al-Qā'im, Iraq." There is no sign of loudspeakers, but the photographs do show the interior of one of the shipping containers behind the disused railway station.

In the first of the photographs, two powerfully built American soldiers are pushing the detainee through a landscape of corrugated iron and barbed wire. He doesn't look hard to push. He is as skinny as a rake. A rag covers his face. One of the soldiers has a handgun pressed to the back of his neck. His finger is on the trigger.

In all the other photographs, the detainee is inside the shipping container. He is barefoot, a thin plastic strap binds his ankles, and he's crouched in the corner, up against the silver corrugated wall. The metal floor is covered with brown dust and pools of liquid. Right at the back of the shipping container, deep in the shadows, you can just make out the figure of another detainee, lying in a huddle on the floor, his face masked by a hood.

Now the rag only covers the first man's eyes, so you can see his face, which is deeply lined, like an old man's, but his wispy mustache reveals that he's probably about seventeen. He's wearing a torn white undershirt, covered with yellow and brown stains. There's an open wound on one of his skinny arms, and above it someone has written a number with a black marker pen.

He might have done terrible things. I know nothing about him other than these seven fragments of his life. But I can say this. In the last photograph he is screaming so hard it almost looks as if he's laughing.

9. THE DARK SIDE

"We don't experiment on each other," Sid Heal had said to me in Los Angeles in early April 2004. "Not in our culture."

A week or two passed. And then the other photographs appeared. They were of Iraqi prisoners in the Abu Ghraib jail on the outskirts of Baghdad. A twenty-one-year-old U.S. reservist named Private Lynndie England had been snapped dragging a naked man across the floor on a leash. In another photograph she stood grinning, a cigarette dangling from her mouth, while she pointed at the genitals of a row of naked, hooded men.

Lynndie England, with her pixie haircut and sweet young face, was the star of many of the photographs. It was she who knelt laughing behind a pile of naked prisoners. They had been forced to build themselves into some kind of human pyramid. Perhaps it was her underwear that was draped over the head of a naked Iraqi who was tied to a metal bed frame, his back arched excruciatingly.

It seemed as if a small group of military guards, with Lynndie England at their center, had used Abu Ghraib to fulfill their sexual fantasies, and that their downfall had been their desire to take trophy photographs.

Defense Secretary Donald Rumsfeld flew to the jail. He told the assembled troops that the events shown in the pictures were the work of "a few who have betrayed our values and sullied the reputation of our country. It was a body blow to me. Those who committed crimes will be dealt with, and the American people will be proud of it, and the Iraqi people will be proud."

The army hung a sign on the gate of the jail that read, AMERICA IS A FRIEND OF ALL THE IRAQI PEOPLE.

Lynndie England was arrested. By then she was back in the United States, five months pregnant, performing desk duties at Fort Bragg. It turned out that she came from a poor town in deepest West Virginia and she had lived for a time in a trailer. For some commentators, that explained everything. DELIVERANCE COMES TO IRAQ ran one headline.

In the 1972 American film *Deliverance,* Bobby the overweight insurance salesman (Ned Beatty) is made to strip. He is then raped from behind by the bigger of the two hillbillies, all the while being forced to squeal like a pig. Maybe it's time to rethink whether these characters were exaggerations. Ms. England definitely does hail from hillbilly country.

The pictures could hardly have been more repulsive, but they were especially so for the people of Iraq, who had long been force-fed Saddam's view that America was, at its heart, uncontrollably depraved and imperialistic. Here were young

Muslim men—captives—being humiliated and overwhelmed by what looked like grotesque U.S. sexual decadence. It struck me as an unhappy coincidence that young Lynndie England and her friends had created a tableau that was the epitome of what would most disgust and repel the Iraqi people, those people whose hearts and minds were the great prize for the coalition forces and also for the Islamic fundamentalists.

But then word got out through Lynndie England's lawyers that her defense was that she had been acting under orders, softening the prisoners up for interrogation, and that the people giving the orders were none other than military intelligence, the unit once commanded by Major General Albert Stubblebine III.

It was sad to remember all that nose banging and cutlery bending, and to think of how General Stubblebine's good intentions had come to this. His soldiers would never have resorted to such terrible acts. They would have instead performed breathtaking psychic feats coupled with remarkable acts of philanthropy.

I called General Stubblebine.

"What was your first thought when you saw the photographs?" I asked him.

"My first thought," he said, "was, 'Oh shit!' "

"What was your second thought?"

" 'Thank God that's not me at the bottom of that pyramid.' "

"What was your third thought?"

"My third thought," said the general, "was 'This was not started by some youngsters down in the trenches. This had to have been driven by the intelligence community.' I told

Rima. I said, 'You watch. This was intelligence.' Yep. Some-one *much* higher in intelligence deliberately designed this, advocated it, directed it, trained people to do it. No doubt about it. And whoever that is, he's in deep hiding right now."

"*Military* intelligence?" I asked. "Your old people?"

"It's a possibility," he said. "My guess is no."

"Who then?"

"The Agency," he said.

"*The* Agency?"

"*The* Agency," he confirmed.

"In conjunction with PsyOps?" I asked.

"I'm sure they had a hand in it," said the general. "Sure. No doubt about it."

There was a silence.

"You know," said General Stubblebine, "if they'd just stuck to Jim Channon's ideas, they wouldn't have needed all that crap."

"By Jim Channon's ideas, do you mean the loud music?" I asked.

"Yeah," said the general.

"So the idea of blasting prisoners with loud music," I said, "*definitely* originated with the First Earth Battalion?"

"Definitely," said the general. "No question. So did the frequencies."

"The frequencies?" I asked.

"Yeah, the frequencies," he said.

"What do the frequencies do?" I asked.

"They disequilibrate people," he said. "There's all kinds of things you can do with the frequencies. Jesus, you can take a frequency and make a guy have diarrhea, make a guy sick to the stomach. I don't understand why they even had to do this crap you saw in the photographs. They should have just *blasted* them with frequencies!"

There was a silence.

"Come to think of it, though," he added, a little ruefully, "I'm not sure what the Geneva convention would say about something like that."

"The loud music and the frequencies?"

"I guess no one's even thought about that," said the general. "That's probably an untested set of waters from a Geneva convention perspective."

On May 12, 2004, Lynndie England gave an interview to Denver-based TV reporter Brian Maas:

BRIAN MAAS: Did things happen in this prison to those Iraqi prisoners worse than what we've seen in these photographs?

LYNNDIE ENGLAND: Yes.

BRIAN MAAS: Can you tell me about that?

LYNNDIE ENGLAND: No.

BRIAN MAAS: What were you thinking when those photographs were taken?

LYNNDIE ENGLAND: I was thinking it was kind of weird. . . . I didn't really, I mean, want to be in any pictures.

BRIAN MAAS: There's a photograph that was taken of you holding an Iraqi prisoner on a leash. How did that come about?

LYNNDIE ENGLAND: I was instructed by persons in higher rank to "stand there, hold this leash, and look at the camera." And they took a picture for PsyOps and that's all I know. . . . I was told to stand there, give the thumbs-up, smile, stand behind all of the naked Iraqis in the pyramid [have my picture taken].

BRIAN MAAS: Who told you to do that?

LYNNDIE ENGLAND: Persons in my higher chain of command . . . They were for PsyOps reasons and the reasons worked. So to us, we were doing our job, which meant we were doing what we were told, and the outcome was what they wanted. They'd come back and they'd look at the pictures and they'd state, "Oh, that's a good tactic, keep it up. That's working. This is working. Keep doing it, it's getting what we need."

Lynndie England seemed to be saying that the photographs were nothing less than an elaborate piece of PsyOps theater. She said that the PsyOps people who told her to "keep doing it, it's getting what we need" did not wear name tags. I was beginning to wonder whether the scenarios had, in fact, been carefully calculated by a PsyOps cultural specialist to present a vision that would most repel young Iraqi men. Could it be that the acts captured in the photographs were not the point at all, and the photographs *themselves* were the thing? Were the photographs intended to be shown only to individual Iraqi prisoners to scare them into cooperating, rather than to get out and scare the whole world?

After I heard the interview with Lynndie England, I dug out my notes of my time at PsyOps. The unit had let me into their Fort Bragg headquarters to show me their CD collection in October 2003, the same month as the Abu Ghraib pictures had been taken. I skimmed through all the talk of "unfulfilled needs" and "desired behavior" until I found my conversation with the friendly boffin in civilian clothing, the "senior cultural analyst" named Dave, who specializes in the Middle East.

Our conversation had at the time seemed innocuous. We were talking about PsyOps "products" in general. All Psy-Ops materials were known as "products"—their radio shows, their leaflets, and so on.

As I reread my notes, what he had said to me took on a whole new resonance.

"We think about how an Iraqi will react to our products, not how an American will react to our products," he said.

He told me they had boards—committees of military analysts and specialists—who look at each product to see whether it furthers the cause of U.S. foreign policy.

"And if it passes muster," he said, "we'll produce it, either here or forward [in Iraq]."

Then Dave spoke about how the target audience for their "products"—Iraqi forces or Iraqi civilians or Iraqi detainees—were not always the most willing customers.

"It's not like selling Coke," he said. "Sometimes you're trying to sell someone something that you know they might not want in their hearts. So it causes ambiguities, and problems. And they have to think about it. It's more like selling someone vitamin D to drink. Something they may not want, but they need it to survive."

"Interesting," I said.

"It causes ambiguities," he said.

10. A THINK TANK

In early 2004 I heard a rumor that Jim Channon had begun privately meeting with General Pete Schoomaker, the new chief of staff of the United States Army.

President Bush had appointed General Schoomaker to the post on August 4, 2003. His "arrival message," to use the military vernacular for an acceptance speech, included the following sentences:

> War is both a physical reality and a state of mind. War is ambiguous, uncertain and unfair. When we are at war, we must think and act differently. We must anticipate the ultimate reality check—combat. We must win both the war and the peace. We must be prepared to question everything. Our soldiers are warriors of character. . . . Our azimuth to the future is good.

Azimuth? I looked it up. It is "the direction of a celestial object." News of General Schoomaker's meetings with Jim Channon did not come as a great surprise to me. (In addition to the linguistic clues, in fact, the timeline of General

Schoomaker's career fit. He had been a commander of Special Forces at Fort Bragg between February 1978 and August 1981, and also in the latter half of 1983, during the periods when the Jedi Warriors and the goat starers were at their most active within his corner of the base. I can't believe he hadn't known about, or indeed sanctioned, their endeavors.)

The rumor was that General Schoomaker was considering bringing Jim back from retirement to create, or contribute to, a new and secret think tank, designed to encourage the army to take their minds further and further outside the mainstream.

Jim had been a member of a similar group back in the early 1980s. It was called Task Force Delta, and it comprised three hundred or so high-ranking soldiers who met four times a year for rituals and brainstorming sessions at Fort Leavenworth, and spent the time in between communicating with one another through something they called the Meta Network, which was an early incarnation of the Internet.

It was a Task Force Delta soldier named Colonel Frank Burns, one of Jim Channon's oldest friends, who launched this technology for the army in the late 1970s. In 1983 Colonel Burns wrote a poem in which he envisioned how his fledgling communications network might one day influence the world:

> Imagine the emergence of a new meta-culture.
> Imagine all kinds of people everywhere
> getting committed to human excellence,
> getting committed to closing the gap
> between the human condition
> and the human potential . . .

And imagine all of us hooked up
with a common high tech communications system.
That's a vision that brings tears to the eyes.
Human excellence is an ideal
that we can embed
into every formal human structure
on our planet.
And that's really why we're going to do this.
And that's also why
The Meta Network is a creation
we can love.

Notwithstanding Colonel Burns's failure to foresee that people would use the Internet mostly to access porn and look themselves up on Google, his prescience was admirable. This same colonel is also widely believed, along with Jim Channon, to be the inspiration behind the recruitment slogan "Be All You Can Be," and the related jingle that all but single-handedly transformed the army's fortunes in the 1980s. Colonel Burns attributed his ideas to reading Jim's *First Earth Battalion Operations Manual.*

A lack of recruits had been the great crisis facing the army back then. It was no wonder, now that General Schoomaker, a Jim Channon fan, was in charge of the army, that these men would be enlisted once again to contribute their ideas to the new crisis, the War on Terror.

Jim e-mailed to say that the General Schoomaker think-tank rumors were true. The idea had come about, he explained, "because Rumsfeld has now openly asked for creative input on the war on terrorism . . . mmmm."

Jim added that he didn't want me to contact General Schoomaker for a comment: "I cannot bear the thought you would interrupt this man's important day with such a gratuitous request. Get a grip! This is media sickness and is grinding the world to a halt! I know you understand."

But Jim did offer some information about his input into George W. Bush's foreign policy:

> The Army has requested my services to teach the most highly selected Majors. The First Earth Battalion is the teaching exemplar of choice. I have done that in the presence of General Pete Schoomaker . . . I am in contact with players who are or have recently been in Afghanistan and Iraq. I have sent in exit strategy plans based on Earth Battalion ideals.
>
> I talk weekly with a member of a stress control battalion in Iraq who carries the manual and uses it to inform his teammates of their potential service contribution. Remember, the battalion mythology operates like folklore. It is passed [on] in stories, not assignments or real world artifacts. The results are ubiquitous, infectious, but not archived well by definition.

Although Jim professed no interest in the "real world artifacts" inspired by him, scattered around the War on Terror, I had become somewhat obsessed with identifying them.

Little pieces of the First Earth Battalion were turning up all over postwar Iraq. A former military spy I spoke to divided Jim's modern-day fans into two categories—the Black Ninjas and the White Ninjas—and that's how I came to see them too.

The 785th Medical Company Combat Stress Control unit, based in Taji, twenty miles north of Baghdad, were White Ninjas. One of their soldiers, Christian Hallman, e-mailed me:

> I utilize many FEB technologies—meditation, yoga, qigong, relaxation, visualization—all part of the FEB toolbox for treating combat stress. It would be great if you came over to Iraq to interview me, but I have to clear it with my commander first. He has read some of the FEB literature that I gave him and has even spoken with Jim over the telephone.

The next day Christian e-mailed me again: "My commander needs to speak with the XO before making his decision." And then, on the third day:

> My commander has declined permission. He doesn't want to run the risk of possibly distorting what we do and our reputation. Sometimes politics wins out.
>   Peace in the Middle East,
>   Christian.

A few weeks after I received this e-mail I learned of a fact that struck me as so bizarre, so incongruous, that I didn't know what to do with it. It was at once banal and extraordinary, and utterly inconsistent with the other facts that encircled it. It was something that had happened to a Mancunian called Jamal al-Harith in a place called the Brown Block. Jamal doesn't know what to make of it either, so he has put it to one side, and mentioned it to me only as an afterthought when I met him in the coffee bar of the Mal-maison Hotel, near Manchester Piccadilly Station, on the morning of June 7, 2004.

Jamal is a web site designer. He lives with his sisters in Moss Side. He is thirty-seven, divorced, with three children. He said he pressumed that MI5 had followed him there to the hotel, but he'd stopped worrying about it. He said he kept seeing the same man watching him from across the street, leaning against a car, and whenever the man thought he'd been spotted he looked briefly panicked and immediately bent down to fiddle casually with his tire.

Jamal laughed when he told me this.

Jamal was born Ronald Fiddler, into a family of second-generation Jamaican immigrants. When he was twenty-three he learned about Islam and converted, changing his name to Jamal al-Harith for no particular reason, other than that he liked the sound of it. He said that al-Harith basically means "seed planter."

In October 2001, Jamal visited Pakistan as a tourist, he said. He was in Quetta, on the Afghanistan border, four days into his trip, when the American bombing campaign began. He quickly decided to leave for Turkey and paid a local truck driver to take him there. The driver said the route would take them through Iran, but somehow they ended up in Afghanistan, where they were stopped by a gang of Taliban supporters. They asked to see Jamal's passport, and he was promptly arrested and thrown in jail on suspicion of being a British spy.

Afghanistan fell to the coalition. The Red Cross visited Jamal in prison. They suggested he cross the border into Pakistan and make his own way back home to Manchester, but Jamal had no money, so instead he asked to be put in contact with the British embassy in Kabul.

Nine days later—while he waited in Kandahar for the embassy to transport him home—the Americans picked him up.

"The Americans," Jamal said, "kidnapped me." When he said "kidnapped" he looked surprised at himself for using such a dramatic word.

The Americans in Kandahar told Jamal he needed to be sent to Cuba for two months for administrative processing, and so on, and the next thing he knew he was on a plane, shackled, his arms chained to his legs, and then chained to a hook on the floor, his face covered in earmuffs and goggles and a surgical mask, bound for Guantanamo Bay.

In the weeks after Jamal's release, two years later, he gave a few interviews, during which he spoke of the shackles and the solitary confinement and the beatings—the things the outside world had already imagined about life inside that mysterious compound. He said they beat his feet with batons, pepper sprayed him, and kept him inside a cage that was open to the elements, with no privacy or protection from the rats and snakes and scorpions that crawled around the base. But these were not sensational revelations.

He spoke with ITV's Martin Bashir, who asked him (off-camera), "Did you see my Michael Jackson documentary?"

Jamal replied, "I've, uh, been in Guantanamo Bay for two years."

When I met Jamal he began to tell me about the more bewildering abuses. Prostitutes were flown in from the

States—he didn't know whether they were there only to smear their menstrual blood on the faces of the more devout detainees. Or perhaps they were brought in to service the soldiers, and some PsyOps boffin—a resident cultural analyst—devised this other job for them as an afterthought, exploiting the resources at the army's disposal.

"One or two of the British guys," Jamal told me, "said to the guards, 'Can *we* have the women?' But the guards said, 'No, no, no. The prostitutes are for the detainees who don't actually *want* them.' They *explained* it to us! 'If you want it, it's not going to work on you.'"

"So what were the prostitutes doing to the detainees?" I asked.

"Just messing about with their genitals," said Jamal. "Stripping off in front of them. Rubbing their breasts in their faces. Not all the guys would speak. They'd just come back from the Brown Block [the interrogation block] and be quiet for days and cry to themselves, so you know something went on but you don't know what. But for the guys who did speak, that's what we heard."

I asked Jamal if he thought that the Americans at Guantanamo were dipping their toes into the waters of exotic interrogation techniques.

"They were doing a lot more than dipping," he replied.

And that's when he told me what happened to him inside the Brown Block.

Jamal said that, being new to torture, he didn't know whether the techniques tested on him were unique to Guantanamo or as old as torture itself, but they seemed pretty weird to him. Jamal's description of life inside the Brown

Block made Guantanamo Bay sound like an experimental interrogation lab, teeming not only with intelligence agents but with ideas. It was as if, for the first time in the soldiers' careers, they had prisoners and a ready-made facility at their disposal, and they couldn't resist putting all their concepts— which had until then languished, sometimes for decades, in the unsatisfactory realm of the theoretical—into practice.

First there were the noises.

"I would describe them as industrial noises," Jamal said. "Screeches and bangs. These would be played across the Brown Block into all the interrogation rooms. You can't describe it. Screeches, bangs, compressed gas. All sorts of things. Jumbled noises."

"Like a fax machine cranking up into use?" I asked.

"No," said Jamal. "Not computer generated. Industrial. Strange noises. And mixed in with it would be something like an electronic piano. Not as in *music*, because there was no rhythm to it."

"Like a synthesizer?"

"Yes, a synthesizer mixed in with industrial noises. All a jumble and a mishmash."

"Did you ever ask them, 'Why are you blasting these strange noises at us?'" I said.

"In Cuba, you learn to accept," said Jamal.

The industrial noises were blasted across the block. But the strangest thing of all happened inside Jamal's own inter-rogation room. The room was furnished with a closed-circuit TV camera and a two-way mirror. Jamal would be brought in for fifteen-hour sessions, during which time they got nothing out of him because, he said, there was nothing to get. He said his past was so clean—not even a parking ticket—that at one point someone wandered over to him and whispered, "Are you an MI5 asset?"

"An MI5 asset!" said Jamal. He whistled. "*Asset!*" he repeated. "That was the word he used!"

The interrogators were getting more and more cross with Jamal's apparent steely refusal to crack. Also, Jamal used his time inside the Brown Block to do stretching exercises, keeping himself sane. Jamal's exercise regime made the interrogators more angry, but instead of beating him, or threatening him, they did something very odd indeed.

A military intelligence officer brought a ghetto blaster into his room. He put it on the floor in the corner. He said, "Here's a great girl band doing Fleetwood Mac songs."

He didn't blast the CD at Jamal. This wasn't sleep deprivation, and it wasn't an attempt to induce the Bucha Effect. Instead, the agent simply put it on at normal volume.

"He put it on," said Jamal, "and he left."

"An all-girl Fleetwood Mac covers band?" I said.

"Yeah," said Jamal.

This sounded to me like the tip of a very strange iceberg.

"And what happened next?" I asked.

"When the CD was finished, he came back into the room and said, 'You might like this.' And he put on Kris Kristofferson's greatest hits. Normal volume. And he left the room again. And then, when that was finished, he came back and said, 'Here's a Matchbox Twenty CD.'"

"Was he doing it for entertainment purposes?" I asked.

"It's interrogation," said Jamal. "I don't think they were trying to entertain me."

"Matchbox Twenty?" I said.

I didn't know much about Matchbox Twenty. My research revealed them to be a four-piece country rock band from Florida, who do not sound particularly abrasive (like Metallica and "Burn Motherfucker Burn!") or irritatingly repetitive (like Barney and "Ya! Ya! Das Is a Mountain"). They sound a bit like REM. The only other occasion when I had heard of Matchbox Twenty was when Adam Piore from *Newsweek* told me that they too had been blasted into the shipping containers in al-Qā'im.

I mentioned this to Jamal and he looked astonished.

"Matchbox Twenty?" he said.

"Their album *More Than You Think You Are*," I said.

There was a silence.

"I thought they were just playing me a CD," said Jamal. "Just playing me a CD. See if I like music or not. Now I've heard this, I'm thinking there must have been something else going on. Now I'm thinking, why did they play that same CD to me as well? They're playing this CD in Iraq and they're playing the same CD in Cuba. It means to me there is a *program*. They're not playing music because they think people like or dislike Matchbox Twenty more than other music. Or Kris Kristofferson more than other music. There is a reason. There's something else going on. Obviously I don't know what it is. But there must be some other intent."

"There must be," I said.

Jamal paused for a moment and then he said, "You don't know how deep the rabbit hole goes, do you? But you know it is deep. You know it is deep."

11. A HAUNTED HOTEL

Joseph Curtis (not his real name) worked the night shift at the Abu Ghraib prison in the autumn of 2003. Then he was exiled by the army to a town in Germany. The threat of a court-martial hung over him. He had given an interview about what he had seen to an international press agency, thus incurring the wrath of his superiors. Even so, against his own better judgment, and against his lawyers' advice, he agreed to meet me, secretly, at an Italian restaurant on a Wednesday in June 2004. I'm not entirely sure why he was willing to risk further censure. Perhaps he felt he couldn't sit back and watch Lynndie England and the other military personnel captured in the photographs be scapegoats just for following orders.

We sat on the balcony of the restaurant and he pushed his food around his plate.

"You ever see *The Shining*?" he said.

"Yes," I said.

"Abu Ghraib was like the Overlook Hotel," he said. "It was *haunted*."

"You mean . . ." I said.

I presumed Joseph meant that the place was full of spooks: intelligence officers—but the look on his face made me realize he didn't.

"It was *haunted*," he said. "It got so dark at night. So dark. Under Saddam, people were dissolved in acid there. Women raped by dogs. Brains splattered all over the walls. This was worse than the Overlook Hotel because it was *real*."

"In *The Shining*," I said, "it was the building that turned Jack Nicholson insane. Was it the building that turned the Americans crazy at Abu Ghraib?"

"It was like the building wanted to be back in business," said Joseph.

Joseph wore a University of Louisiana athletics department T-shirt. He had the U.S. soldier's jarhead haircut—shaved at the sides, a short crop on top. He said he couldn't believe how much money was floating around the army these days. These were the golden days, in budgetary terms. One day he had taken his truck in for repairs, and the soldier who looked it over had said, "You need new seats."

Joseph said it didn't look like the seats needed replacing.

The soldier replied that they had two hundred thousand dollars in their budget and if they didn't spend it by the end of the month they had to give it back.

"So," the soldier slowly repeated, "you need new seats."

Joseph said I wouldn't believe how many plasma screens there were in Iraq, for teleconferencing purposes and so on. They'd had perfectly good TVs, but trucks full of plasma screens just arrived one day, because that's how much money was floating around.

In January 2004, the influential think tank and lobbying group GlobalSecurity revealed that George W. Bush's government had filtered more money into their Black Budget than any other administration in American history.

The amount of money an administration spends on its Black Budget can be seen as a tantalizing barometer of its proclivity toward weirdness. Black Budgets often just fund Black Ops—highly sensitive and deeply shady projects such as assassination squads, and so on, which remain secret not only to protect the Black Operators but to protect Americans, who generally don't want to think about such things. But Black Budgets also fund investigations into schemes so bizarre that their disclosure might lead voters to believe that their leaders have taken leave of their senses. George W. Bush's administration had, by January 2004, channeled approximately $30 billion into the Black Budget—to be spent on God knows what.

I had to strain to hear Joseph over the late-night roadworks as he told me about the darkness at Abu Ghraib, and how that darkness led to "the beast in man really coming out there" and the never-ending, bountiful budget.

"Abu Ghraib was a tourist attraction," he said. "I remember one time I was woken up by two captains. 'Where's the death chamber?' They wanted to see the rope and the lever. When Rumsfeld came to visit, he didn't want to talk to the soldiers. All he wanted to see was the death chamber."

Joseph took a bite of his food.

"Yeah, the beast in man really came out at Abu Ghraib," he said.

"You mean in the photographs?" I asked.

"Everywhere," he said. "The senior leadership were screwing around with the lower ranks. . . ."

I told Joseph I didn't understand what he meant.

He said, "The senior leaders were having sex with the lower ranks. The detainees were raping each other."

"Did you ever see any ghosts?" I asked him.

He stopped eating and pushed his food around the plate again.

"There was a darkness about the place," he replied. "You just got this feeling there was always something there, lurking behind you in the darkness, and that it was very mad."

I asked Joseph if there was anything *good* at Abu Ghraib, and he paused, then said it was good that Amazon.com delivered there. Then he remembered something else that was good. He said there was a genius at making model planes there. He made them out of old ration boxes, and he hung them from the ceiling in the isolation block. One time, Joseph said, someone came up to him and said, "You've *got* to see these model planes! They're incredible! One of the guards in the isolation block has hung a bunch of them on the ceiling. Hey, and while you're there, you can take a look at the high values!"

The "high values" were what the U.S. army called the suspected terrorists, insurgent leaders, rapists, and child molesters, although things were so out of control in postwar Iraq that many of the high values might have just been passersby picked up at checkpoints because the soldiers didn't like the look of them.

Joseph was in charge of the superclassified computer network at Abu Ghraib. He had set up the system and handed out the user names and passwords. His job didn't take him into the isolation block, even though it was just down the corridor. So he accepted the invitation. He got up from behind his desk and walked toward the model planes and the high values.

A few weeks before I met Joseph, it was revealed, by Seymour Hersh in *The New Yorker*, that on April 9, 2004, Specialist Matthew Wisdom told an Article 32 hearing (the military equivalent of a grand jury): "I saw two naked detainees [in the isolation block at Abu Ghraib], one masturbating to another kneeling with its mouth open. I thought I should just get out of there. I didn't think it was right. . . . I saw SSG [Ivan] Frederick walk toward me, and he said, 'Look what these animals do when you leave them alone for two seconds.' I heard PFC [Lynndie] England shout out, 'He's getting hard.'"

The isolation block was where all the photographs were taken—Lynndie dragging a naked man across the floor on a leash, and so on.

Joseph turned the corner into the isolation block.

"There were two MPs there," he told me. "And they were constantly screaming. 'SHUT THE FUCK UP!' They were screaming at some old guy, making him repeat a number over and over: 156403. 156403. 156403.

"The old guy couldn't speak English. He couldn't pronounce the numbers.

"'I CAN'T FUCKING HEAR YOU.'

"The guy repeated the number, twice.

"'LOUDER. FUCKING LOUDER.'

"Then they saw me. 'Hey, Joseph! How are you? I CAN'T FUCKING HEAR YOU. LOUDER.'

"I repeated the numbers, twice."

Joseph said that the MPs had basically gone straight from McDonald's to Abu Ghraib. They knew nothing. And now they were getting scapegoated because they happened to be identifiable in the photographs. They just did what the military intelligence people, Joseph's people, had told them to do. PsyOps was just a phone call away, Joseph said. And the military intelligence people all had PsyOps training anyway. The thing I had to remember about military intelligence was that they were the "nerdy-type guys at school. You know. The outcasts. Couple all that with ego, and a poster on the wall saying BY CG APPROVAL (Commanding General Approval), and suddenly you have guys who think they govern the world. That's what one of them said to me. 'We govern the world.'"

"Were there many intelligence officers at Abu Ghraib?" I asked Joseph.

"There were intelligence people turning up there I never even knew *existed,*" he said. "There was a unit from Utah. All *Mormons*. It was a real casserole of intelligence, and they all had to come to me to get their user names and passwords. They were from all kinds of units, civilian guys, and translators. Two British guys showed up. They were older, in uniform, and were getting themselves properly installed. They had laptops and a desk."

An aide to Condoleezza Rice, the White House national security adviser, visited the prison also, to inform the interrogators sternly that they weren't getting useful enough information from the detainees.

"Then," Joseph said, "a whole platoon of Guantanamo people arrived. The word got around. 'Oh God, the Gitmo guys are here.' *Bam!* There they were. They took the place over."

Perhaps Guantanamo Bay was Experimental Lab Mark 1, and whatever esoteric techniques worked there were exported to Abu Ghraib. I asked Joseph if he knew anything about the music. He said, sure, they blasted loud music at the detainees all the time.

"What about quieter music?" I asked, and I told him Jamal's story about the ghetto blaster and the Fleetwood Mac all-girl covers band and Matchbox Twenty.

Joseph laughed. He shook his head in wonderment.

"They were probably fucking with his head," he said.

"You mean they did it just *because* it seemed so weird?" I asked. "The incongruity was the point of it?"

"Yeah," he said.

"But that doesn't make sense," I said. "I can imagine that might work on a devout Muslim from an Arab country, but Jamal is British. He was raised in Manchester. He knows all about ghetto blasters and Fleetwood Mac and country-and-western music."

"Hm," said Joseph.

"Do you think . . . ?" I said.

Joseph finished my sentence for me.

"Subliminal messages?" he said.

"Or something like that," I said. "Something *underneath* the music."

"You know," said Joseph, "on a surface level that would be ridiculous. But Guantanamo and Abu Ghraib were *anything* but surface."

12. THE FREQUENCIES

Perhaps, I thought, one way of solving this mystery was to follow the patents, follow them like a tracker follows footprints in the snow, and then, like in a horror film, see how the footprints vanish. Was there, somewhere out there, a paper trail of patents for subliminal sound technologies, or frequency technologies, that simply vanished into the classified world of the United States government?

Yes. There was. And the inventor in question was a mysterious figure named Dr. Oliver Lowery.

On October 27, 1992, Dr. Oliver Lowery, of Georgia, was the recipient of U.S. Patent #5,159,703. His invention was something he called a Silent Subliminal Presentation System:

A silent communications system in which non-aural carriers, in the very low or very high audio-frequency range or in the adjacent ultrasonic frequency spectrum are amplitude or frequency modulated with the desired intelligence and propagated acoustically or vibrationally, for inducement into the brain, typically through the use of loudspeakers, earphones, or piezoelectric transducers. The

modulated carriers may be transmitted directly in real time
or may be conveniently recorded and stored on mechani-
cal, magnetic or optical media for delayed or repeated
transmission to the listener.

The publicity material that accompanied this patent put it
into plainer language. Dr. Lowery had invented a way in
which subliminal sounds could be put onto a CD where they
would "silently induce and change the emotional state in a
human being."

The following emotional states could, according to Dr.
Lowery, be induced by his invention:

Positive emotions: contentment, duty, faith, friendship,
hope, innocence, joy, love, pride, respect, self-love, and wor-
ship.

Negative emotions: anger, anguish, anxiety, contempt, des-
pair, dread, embarrassment, envy, fear, frustration, grief, guilt,
hate, indifference, indignation, jealousy, pity, rage, regret,
remorse, resentment, sadness, shame, spite, terror, and vanity.

Twelve positive emotions, twenty-six negative ones.

Four years later, on December 13, 1996, Dr. Lowery's company, Silent Sounds Inc., posted the following message on their
web site: "All schematics have [now] been classified by the
U.S. Government and we are not allowed to reveal the exact
details . . . we make tapes and CDs for the German Government, even the former Soviet Union countries! All with the
permission of the U.S. State Department, of course. . . . The
system was used throughout Operation Desert Storm (Iraq)
quite successfully."

For weeks on end I repeatedly telephoned the number I

had found for Dr. Oliver Lowery—it was a Georgia area code, somewhere on the outskirts of Atlanta—but nobody picked up the phone.

Until, one day, someone did.

"Hello?" said the voice.

"Dr. Lowery?" I said.

"I'd prefer you didn't call me that," he said.

"What can I call you?" I said.

"Call me Bud," he said.

I could almost hear him smile down the phone.

"Call me Hamish McLaren," he said then.

I told Hamish/Bud/Dr. Oliver Lowery what I was doing, and he, in return, whoever he was, told me something about his life. He said he was seventy-seven years old, a Second World War veteran, a former Hughes aerospace engineer, and he had endured numerous operations, heart bypasses, and so on. Then he said, "You're the first journalist to find us in four years."

"Find 'us'?" I said.

"You think you're talking to *Georgia*?" he said.

There was a faint mocking tone to his voice.

"I'm sorry?" I said.

He laughed.

"I phoned a Georgia area code," I said.

There seemed to be voices in the background, a lot of commotion, as if Oliver/Bud/Hamish was speaking from the middle of a busy office.

"You'll never be able to print what I am about to tell you," he said, "because there is no way you'll be able to prove that this conversation ever took place."

"So I'm not talking to someone in Georgia?" I said.

"You are talking to someone in a lab where there are numerous PhDs from sixteen countries, including Brits, and the lab is in a fourteen-story building behind three layers of barbed-wire defenses that sure as hell isn't Georgia," he said.

There was a long pause.

"So you're using call divert?" I said, weakly.

I had no idea if this was true. Maybe he was a fantasist, or perhaps he was playing with me for the fun of it, but, as I say, there seemed to be many voices in the background. (Maybe he was just putting those voices in my head.)

The man said that the U.S. military has been researching silent-sound technologies for twenty-five years. He likened this "massive" research to the Manhattan Project.

He said there were good silent sounds—"children exposed to the good sounds in the womb turn out remarkably bright"—and bad silent sounds.

"We only use the bad stuff on the bad guys," he said.

He said the Americans used bad subliminal sounds on Iraqi soldiers during the first Gulf War ("We warped their brains for a hundred days,") but they have had "serious problems getting the subliminally implanted fears out of their heads" during the years that followed.

"Negative things are a devil to get out." He chuckled.

He said ITN news once broadcast a story on the use of silent sounds in the first Gulf War.

(ITN later told me, categorically, that they had never run such a story. Nowhere within their archive database could I find anything approaching it.)

He said, "You can transmit silent sounds into people's

heads via a window in the same way you can fire a laser beam through a window to eavesdrop on a conversation. Conversely, the sounds can be transmitted through the crummiest media—a satellite telephone or a crappy old tape recorder or a ghetto blaster."

He said New Scotland Yard uses the technology, but he wouldn't tell me how. He said the Russians use it too. And that was it. He cut the conversation short. He wished me the best and he hung up and I was left reeling and entirely unsure of everything he had said.

This man seemed to have verified one of the world's most enduring and least plausible conspiracy theories. For me, the idea that the government would surreptitiously zap heads with subliminal sounds and remotely alter moods was on a par with the idea that they were concealing UFOs in military hangars and transforming themselves into twelve-foot lizards. This conspiracy theory has persisted because it contains all the crucial ingredients—the hidden hand of big government teaming up with Machiavellian scientists to take over our minds like body snatchers.

The thing is, in this context, Jamal's Fleetwood Mac all-girl-covers-band-ghetto-blaster experience inside the Brown Block at Guantanamo Bay suddenly made sense.

Jamal seemed fine when I met him in Manchester. I asked him if he felt at all unusual after listening to Matchbox Twenty and he said no. But one shouldn't read too much into this. There is a very strong chance, given the history of the goat staring and the wall walking and so on, that they blasted Jamal with silent sounds and it just didn't work.

There was one thing I could chase up. Dr. Oliver Lowery (or whoever he was) had mentioned to me a Dr. Igor Smirnov. He said Igor Smirnov had undertaken similar U.S. government work in the field of silent sounds. I looked Dr. Smirnov up. I found him in Moscow. I corresponded with his office, and his assistant (Dr. Smirnov speaks little English) told me the following curious story.

It is a story the FBI has never denied.

Igor Smirnov was not prospering in the post-Cold War Moscow of 1993. His finances were so bleak that when the Russian mafia turned up at his laboratory one evening, pressed the bell marked, somewhat ominously, "Institute for Psycho-Correction," and told Igor they'd pay him handsomely if he could subliminally influence certain unwilling businessmen to sign certain contracts, he almost accepted their offer. But in the end it seemed just too frightening and unethical and he turned the gangsters down. His regular clients—the schizophrenics and the drug addicts—may have been poor payers but at least they weren't the mafia.

Igor's day-to-day work in the early 1990s was something like this: A heroin addict would turn up at his lab very upset because he was a father-to-be but try as he might he cared more about the heroin than his unborn child. So he'd lie on a bed, and Igor would blast him with subliminal messages. He'd flash them onto a screen in front of the addict's eyes and blast them through earphones, disguised by white noise, and the messages would say "Be a good father. Fatherhood is more important than heroin." And so on.

This was a man once fêted by the Soviet government,

which—ten years earlier—had instructed him to blast his silent messages at Red Army troops on their way to Afghanistan. Those messages said, "Do not get drunk before battle."

But the glory days were long gone by March 1993—the month Igor Smirnov received a telephone call, out of the blue, from the FBI. Could he fly to Arlington, Virginia, right away? Igor Smirnov was intrigued, and quite amazed, and he got on a plane.

The U.S. intelligence community had been spying on Igor Smirnov for years. It seemed he'd succeeded in creating a system of influencing people from afar—putting voices into their heads, remotely altering their outlook on life—perhaps without the subjects even knowing it was being done to them. This was a tangible, real-life, mechanistic version of General Wickham's prayer groups, or Guy Savelli's goat staring, the kind of system the ambient composer Steven Halpern had suggested to Jim Channon back in the late 1970s. The question was: Could Igor do it to David Koresh?

Could he put the voice of God into David Koresh's head?

The Branch Davidians, an offshoot of the Seventh Day Adventists, had been living around Waco, Texas, predicting an imminent Judgment Day, since 1935. When Vernon Howell took over the church's leadership in the late 1980s, and declared himself a Christ-figure, the anointed one, the seventh and final messenger as outlined in the Book of Revelation, and changed his name to David Koresh and started selling weapons illegally to fund his congregation's separatist lifestyle, the Bureau of Alcohol, Tobacco and Firearms began to take an interest. They imagined that a high-profile raid on the church would be good for agency morale and PR. So they tipped off the local media, told them the Branch Davidians were theologically incomprehensible, nuts, and heavily armed (they were, but basically in the way that gun shops are heavily armed), and they were going in.

What the BATF failed to predict was that Koresh had been waiting for a confrontation like this, and relished the prospect. It was his destiny to be attacked by a hostile army representing an out-of-control, sinister, heavy-handed, new-world-order-type Babylon government.

On February 28, 1993, a hundred or so BATF agents stormed the church, but the raid turned into a gun battle, during which four agents were killed, and the gun battle turned into a siege.

There is something far too familiar, in retrospect, about the whole thing. At Waco, just as at Abu Ghraib, the U.S. government behaved like a grotesque caricature of itself. The anti-big-government American right wing had harbored paranoid fantasies about the Clinton administration heavy-handedly destroying the lives of simple people who wanted to live free, and Waco was the place where those conspiracy theories came true. Much of the Iraqi population had been fed similarly wild conspiracy theories about American imperialistic hedonism—that the United States was violently out of control and determined to force its corruption and decadence on the devout—and Abu Ghraib was the place where those conspiracy theories came true.

But there is a more disturbing parallel. David Koresh's

Branch Davidians also seem to have been considered guinea pigs in the middle of a long-awaited golden opportunity, an opportunity to try stuff out.

Back in 1993, the problem for advocates of out-of-the-box thinking within the U.S. government and military establishments was that there was nobody suitably wicked out there on whom to test their ideas. The outlook was so hopeful, in fact, that a State Department social scientist named Francis Fukuyama declared in 1989, to widespread international acclaim, that it was the end of history. Western democratic capitalism had proved so superior to all its historical rivals, wrote Fukuyama, that it was finding acceptance everywhere in the world. There was simply nothing nefarious out there on the horizon. Although this eventually turned out to be just about the worst prediction ever, in 1993 it seemed too real. These were the fallow years for those who wanted to experiment with new ideas on suitable adversaries.

And then the Waco siege came along.

First there were the noises. Midway through the siege—in the middle of March 1993—the sounds of Tibetan Buddhist chants, screeching bagpipes, crying seagulls, helicopter rotor blades, dentist drills, sirens, dying rabbits, a train, and Nancy Sinatra's "These Boots Are Made for Walking" began to blast into the church. It was the FBI, in this instance, who did the blasting. There were seventy-nine members of David Koresh's congregation in there, including twenty-five children (twenty-seven if you count the unborn ones). Some of the parishioners put cotton wool in their ears, a luxury that was later unavailable to Jamal at Guantanamo and the prisoners inside the shipping containers in al-Qā'im. Others apparently tried to enjoy it by ironically pretending it was a disco. This wasn't easy, as one of them, Clive Doyle, told me when I telephoned him.

Clive Doyle is one of the very few survivors of the fire that ended the siege.

"Very rarely did they play a song straight through," he said. "They distorted it by slowing it down or speeding it up. And the Tibetan monks were pretty ominous." Then, apropos of nothing, he said, "Do you think they blasted us with those subliminal sounds?"

"I don't know," I said. "Do you?"

"I don't know," he said. "We figured they were experimenting in a lot of different areas. They had a robot that came down the drive one day, with a big antenna sticking out the top. What was that about?"

"I don't know," I said.

"Sometimes," said Clive Doyle, "I think that the FBI were just like idiots, and it was just chaos out there."

It did seem somewhat chaotic. Some of the noises blasted at the Branch Davidians, like the Buddhist chants, I have learned, came from the wife of an agent present at the scene. She worked in a local museum. She just scooped up the tapes and gave them to her husband. The dying rabbit noises were an exception. They came from an FBI agent who used the tape, under normal circumstances, to ferret out coyotes during his regular hunting trips. Furthermore, the FBI carried on blasting the Buddhist chants even after the Dalai Lama had written a letter of complaint because the agent in charge of the speaker system "didn't have anything else to do at night."

My guess is that, just like at Abu Ghraib, there was a

"casserole of intelligence" present, each with their own idea about how to direct the siege. Some of the ideas were inspired by Jim Channon, or inspired by people who were inspired by him. Others were more random. The FBI negotiators taped their telephone conversations with David Koresh and his deputies. Excerpts from these tapes illustrate two things: the people within the church were, somewhat alarmingly, of one mind—David Koresh's mind; the people on the outside of the church were, even more alarmingly, of no cohesive mind-set whatsoever.

STEVE SCHNEIDER (Branch Davidian): Who's controlling these guys? You've got guys out there right now pulling their pants down, men that are mature, sticking their butts in the air and flipping the finger.

FBI NEGOTIATOR: Uh. Give me a moment. The guys that gravitate towards riding in tanks, jumping out of airplanes, have a little different mind-set from you and I, would you agree?

STEVE SCHNEIDER: I agree with you. But somebody's gotta be above these guys.

FBI NEGOTIATOR: Sure.

JIM CAVANAUGH (FBI negotiator): I think we need to set the record straight. There were no guns on those helicopters.

DAVID KORESH: That's a lie. That is a lie. Now, Jim, you're a damn liar. Let's get real.

189

JIM CAVANAUGH: David, I—

DAVID KORESH: No, you listen to me. You're sitting there telling me that there were no guns on that helicopter?

JIM CAVANAUGH: I said they didn't shoot.

DAVID KORESH: You are a damn liar.

JIM CAVANAUGH: Well, you're wrong, David.

DAVID KORESH: You are a liar.

JIM CAVANAUGH: Okay. Well, just calm down. . . .

DAVID KORESH: No! Let me tell you something. That may be what you might want the media to believe, but there's other people that saw too. Now tell me, Jim, you're honestly going to say those helicopters didn't fire on any of us?

JIM CAVANAUGH (after a long silence): David?

DAVID KORESH: I'm here.

JIM CAVANAUGH: Uh, yeah, uh, what I'm saying is those helicopters did not have mounted guns. Okay? I'm not disputing the fact that there might have been fire from the helicopters. Do you understand what I'm saying?

DAVID KORESH: Uh, no.

UNIDENTIFIED LITTLE GIRL: Are they going to come in and kill me?

UNIDENTIFIED NEGOTIATOR: No. Nobody's coming. Nobody's coming.

And this, from a press conference that occurred midway through the siege:

JOURNALIST: Mr. Ricks, is there a consideration to use psychological warfare? Have you discussed it at all?

BOB RICKS (FBI spokesman): I don't know what psychological warfare is.

JOURNALIST: It was reported in the paper that you would play loud music, put bright lights on the compound all night, to try to agitate the entire group. Is that possible?

BOB RICKS: We will not discuss tactics of that sort, but I would say the chances of doing that sort of activity are minimal.

I have met Bob Ricks. He has been one of the FBI's most outspoken critics of the Waco siege, and he all but single-handedly prevented a similar raid from occurring on a group of white supremacists in northern Oklahoma, at a place called Elohim City. I don't think that Bob Ricks was lying during the press conference. I think the FBI's left hand didn't know what its right hand was up to.

At Waco, just like at Abu Ghraib, the Jim Channon–like thinkers seem to have had to bide their time, wait for the finger flippers and the helicopter snipers to have their turn.

My guess is that the musical bombardment was inspired by a similar event that occurred four years earlier in Panama City. The battle between General Stubblebine and General Manuel Noriega had long been fought like two wizards standing atop mountains throwing thunderbolts at each other. General Stubblebine had set his psychic spies onto Noriega, who had countered by inserting little slips of paper into his shoes, and so on.

In the end, Noriega turned up at the Vatican embassy in Panama City, and PsyOps arrived on the scene with loud-speakers attached to their trucks, which were used to repeat-edly blast the building with Guns N' Roses' "Welcome to the Jungle." If this event was inspired (directly or indirectly) by Jim Channon's manual, it is appropriate that Noriega—who had given General Stubblebine so much hassle that he couldn't concentrate on walking through his wall—was finally felled by another First Earth Battalion idea.

I telephoned a dozen witnesses to the siege at Waco—jour-nalists and intelligence agents—and I asked them if they knew of any strange goings-on beyond the music and the robot with the antenna. Three of them told me the same story. I can't prove it, so it remains a rumor—one that sounds plausible but, then again, entirely implausible.

The rumor I heard involves a man I'll call Mr. B. He enlist-ed in the U.S. military in 1972, and between 1973 and 1989 he was in the Special Forces unit at Fort Bragg, where he took part in the various General Stubblebine inspired supersoldier programs. As a result he became—in the words of one man I spoke to—"not only the greatest surreptitious break-in guy in the armed forces, but in the whole government."

Mr. B. could break in anywhere, unseen and unheard. He had, to all intents and purposes, wholly and extraordinarily mastered level three of Glenn Wheaton's Jedi Warrior code: invisibility. But Mr. B. used his powers for bad. He was convicted, in 1989, of breaking into women's apartments and raping them. He was sentenced to life imprisonment.

A soldier I cannot name swears that, on April 18, 1993, he spotted Mr. B. surreptitiously entering David Koresh's church. Perhaps his four years in prison had diminished his powers, for the soldier recognized him immediately. He said nothing at the time, because he knew he had witnessed a Black Op. An intelligence agency must have sprung Mr. B. from jail.

The rumor ends like this: Mr. B. entered the Koresh compound, checked that the bugging devices were in good working order, fixed those that weren't, crept out again, was transported back to his jail cell in Colorado, and found God. He refused to grant me an interview because he said he no longer wanted to dwell on his past.

He remains to this day in a maximum-security prison.

That story remains a rumor, whereas Dr. Igor Smirnov's involvement in the Waco siege can safely be considered true.

The FBI flew Dr. Smirnov from Moscow to Arlington, Virginia, where he found himself in a conference room with representatives of the FBI, the CIA, the Defense Intelligence Agency, and the Advance Research Projects Research Agency.

The idea, the agents explained, was to use the telephone lines. The FBI negotiators would bargain with Koresh as usual but, underneath, the silent voice of God would tell Koresh whatever the FBI wanted God to say.

Dr. Smirnov said this was possible.

But then bureaucracy crept into the negotiations. An FBI agent said he was concerned that the endeavor might somehow lead to the Branch Davidians' committing mass suicide. Would Dr. Smirnov sign something to the effect that if they did kill themselves as a result of the voice of God being subliminally implanted in their heads, he would take responsibility?

Dr. Smirnov said he wouldn't sign something like that.

And so the meeting broke up.

An agent told Dr. Smirnov it was a shame it didn't work out. He said they had already co-opted someone to play the voice of God.

Had Dr. Smirnov's technology been put into practice at Waco, the agent said, God would have been played by Charlton Heston.

I was passing through Georgia, and thinking a lot about my telephone conversation with Dr. Oliver Lowery, and so I decided to drive past the address I had for him. It was somewhere in the suburbs of Atlanta. I wondered if I'd find a normal house or something like a fourteen-story building behind three layers of barbed-wire defenses. A gale was blowing so strongly I thought it would tip over my car.

It was a normal, slightly dilapidated wooden house on a middle-class leafy street, and the leaves were swirling around so violently I had to switch on my windshield wipers.

I parked the car and walked up his drive, shielding myself against the gale. I was feeling quite nervous. I knocked. The whole thing happened so fast I can't even describe the person who opened the door. I have an impression of a craggy-looking man in his seventies, his white hair blowing in the wind.

I said, "I'm really sorry to just turn up at your house. If you remember we—"

He said, "I hope the wind doesn't blow you over on your journey back to your car."

And then he closed the door on me.

I walked back down his drive. And then I heard his voice again. I turned around. He was shouting something through a crack in his door. He was shouting, "I hope the wind doesn't blow you away."

I smiled uneasily.

"You'd better take care," he shouted.

13. SOME ILLUSTRATIONS

The shipping container behind the disused railway station in al-Qā'im, Iraq, in which the "I Love You" song from Barney the Purple Dinosaur was played through a loudspeaker to detainees.

In late June 2004 I sent an e-mail to Jim Channon and everyone else I had met during my two-and-a-half-year journey who might have some inside knowledge about the current use of the kinds of psychological interrogation techniques that had first been suggested in Jim's manual. I wrote:

Dear————

    I hope you are well.

    I was talking with one of the British Guantanamo detainees (innocent—he was released) and he told me a very strange story. He said at one point during the interrogations the MI officers left him in a room—for hours and hours—with a ghetto blaster. They played him a series of CDs—Fleetwood Mac, Kris Kristofferson, etc. They didn't blast them at him. They just played them at normal volume. Now, as this man is Western, I'm sure they weren't trying to freak him out by introducing him to Western music. Which leads me to think . . .

    . . . Frequencies? Subliminal messages?

    What's your view on this? Do you know any time when frequencies or subliminal sounds have been used by the U.S. military for sure?

    With best wishes,

    Jon Ronson

I received three replies straightaway:

COMMANDER SID HEAL (the Los Angeles County Sheriff's Department nonlethals expert who told me about the Bucha Effect): Most interesting, but I haven't a clue. I know that subliminal messages can be incorporated and that they have a powerful influence. There are laws prohibiting it in the U.S., but I'm not aware of any uses like you describe. I would imagine, however, that it would be classified and no one without a "need to know" would be aware anyway. If it were frequencies, it would probably need to be in the audible range or they wouldn't need to mask them with other sounds.

SKIP ATWATER: (General Stubblebine's former psychic spying headhunter): You can bet this activity was purposeful. If you can get anybody to talk to you about this, it would be interesting to know the "success rate" of this technique.

JIM CHANNON: Strikes me the story you tell is just plain kindness (which still exists).

I couldn't decide if Jim was being delighfully naive, infuriatingly naive, or sophisticatedly evasive. (Major Ed Dames, the mysterious Art Bell psychic-unit whistleblower and a neighbor of Jim's, had once described him to me in an unexpected way. He'd said, "Don't be taken in by Jim's hippie demeanor. Jim isn't airy at all. He's like the local warlord. He *runs* that part of Hawaii. Jim is a very shrewd man.")

Then Colonel John Alexander responded to my e-mail. Colonel Alexander remains the army's leading pioneer of nonlethal technologies, a role he created for himself in part after reading and being inspired by Jim's manual.

COLONEL ALEXANDER: Re your assertion he was innocent. If so, how did he get captured in Afghanistan? Don't think there were many British tourists who happened to be traveling there when our forces arrived. Or maybe he was a cultural anthropologist studying the progressive social order of

the Taliban as part of his doctoral dissertation and was just
mistakenly detained from his education. Perhaps if you
believe this man's story you'd also be interested in buying a
bridge from me? As for the music, I have no idea what that
might be about. Guess hard rockers might take that as cruel
and unusual punishment and want to report it to Amnesty
International as proof of torture.

Jokes about the use of music in interrogation didn't seem
that funny anymore—not to me, and I doubt they did to him
either. Colonel Alexander has spent a lifetime in the world of
plausible deniability and I think he's got to the stage where
he just trots these things out. Colonel Alexander has just
returned from four months in Afghanistan advising the army
on something he wouldn't talk about.

I e-mailed him back:

Is there anything you can tell me about the use of subliminal
sounds and frequencies in the military's arsenal? If anyone
alive today is equipped to answer that question, surely you are.

His response arrived instantly. He said my assertion that
the U.S. army would ever entertain the possibility of using
subliminal sounds or frequencies "just doesn't make sense."

Which was strange.

I dug out an interview I'd conducted with the colonel the
preceding summer. I hadn't been that interested in acoustic
weapons at that point—I was trying to find out about Sticky
Foam and goat staring—but the conversation had, I now
remembered, briefly touched on them.

"Has the army ever blasted anyone with subliminal sounds?" I had asked him.

"I have no idea," he said.

"What's a 'psycho-correction' device?" I asked him.

"I have no idea," he said. "It has no basis in reality."

"What are silent sounds?" I asked.

"I have no idea," he said. "It sounds like an oxymoron to me."

He gave me a hard look, which seemed to suggest that I was masquerading as a journalist but was, in fact, a dangerous and irrational conspiracy nut.

"What's your name again?" he said.

I felt myself blush. I was suddenly finding Colonel Alexander quite frightening. Jim Channon has a page in his manual dedicated to the facial expression a Warrior Monk should adopt when meeting an enemy or a stranger for the first time. "A consistent, subtle and haunting smile," Jim wrote. "A deep and unblinking look indicating the real person is home and comfortable with all. A quiet and calm stare indicating a willingness to be open." Colonel Alexander was now giving me what I can only describe as a haunting and unblinking stare.

I told him my name again.

He said, "Pixie dust."

"Sorry?" I said.

"This is not something that has been brought up or addressed, and we have covered the waterfront of nonlethal technologies," he said. "We are not warping people's brains or monitoring people or da da da da da. It's just nonsense."

"I'm confused," I said. "I don't know much about this

subject but I'm sure I've seen your name linked with something called a 'psycho-correction device.'"

"It makes no sense," he said. He looked baffled. Then he said that yes, he had sat in on meetings where this sort of thing was discussed, but there was no evidence that machines like this would ever work. "How would you do that [blast someone with silent sounds] without it affecting us? Anybody who's out there would hear it."

"Earplugs?" I said.

"Oh, come on," he said.

"Of course," I said. "You're right."

And then the conversation had moved on to the subject of staring goats to death—"In a scientifically controlled environment," said Colonel Alexander—and that's when he told me that the man who achieved this feat was not Michael Echanis but Guy Savelli.

How could you blast someone with silent sounds "without it affecting us?"

This struck me at the time as an unassailable argument, and one that cut through all the paranoid theories circulating on the Internet about mind-control machines putting voices into people's heads. Of course it couldn't work. In fact, it was a relief to believe Colonel Alexander because it made me feel sensible again, and not the nut his look had suggested I was. Now we were once again two sensible people—a colonel and a journalist—discussing rational things in a sagacious manner.

The thing is, I now realized, if silent sounds *had* been used against Jamal inside an interrogation room at Guantanamo

Bay, there was a clue in Jamal's account, a clue that suggested that military intelligence had craftily solved the vexing problem highlighted by Colonel Alexander.

"He put the CD in," Jamal had said, "and he left the room."

Next, I dug out the recently leaked military report titled *Non-Lethal Weapons: Terms and References*. There were a total of twenty-one acoustic weapons listed, in various stages of development, including the Infrasound ("Very low-frequency sound which can travel long distances and easily penetrate most buildings and vehicles . . . biophysical effects: nausea, loss of bowels, disorientation, vomiting, potential internal organ damage or death may occur. Superior to ultrasound . . .").

And then the last but one entry—the Psycho-Correction Device, which "involves influencing subjects visually or aurally with embedded subliminal messages."

I turned to the front page. And there it was. The coauthor of this document was Colonel John Alexander.

And so our e-mails continued.

I asked for the colonel's permission to include in this book his views on the Guantanamo story, and he replied:

Not sure what you mean by the Guantanamo story. My take on this whole thing is much bigger. IMHO [in my humble opinion] World War X is on, and it is religious. We are now faced with a problem of how to handle prisoners caught in a war that never ends. Nobody has asked that before. The traditional response (over millennia) is to kill them or put them into slavery. Tough to do in today's environment.

It seemed obvious to me what his alternative was, knowing what I did about his area of expertise. If you couldn't kill your adversaries, or keep them imprisoned forever, there was surely only one option left in the Colonel Alexander canon: You change their minds.

The *First Earth Battalion Operations Manual* had encouraged the development of devices that could "direct energy into crowds." History seems to show that whenever there is a great American crisis—the War on Terror, the trauma of Vietnam and its aftermath, the Cold War—its military intelligence is drawn to the idea of thought control. They come up with all manner of harebrained schemes to try out, and they all sound funny until the schemes are actually implemented.

I e-mailed Colonel Alexander to ask if he was indeed advocating the use of some kind of mind-control machine and he replied, somewhat ruefully, and a little guardedly: "If we do go to scrambling minds, then the whole mind-control conspiracy issues arise."

What he meant was, MK-ULTRA.

It was really just about the worst PR the U.S. intelligence services had ever suffered, certainly until the Abu Ghraib photographs came along in 2004. Jim Channon might have pretty much single-handedly invented the idea of the army thinking outside the box (as one of his admirers had once told me), but the CIA had been there before him.

Everyone was still bruised over MK-ULTRA.

14. THE 1953 HOUSE

There is a house in Frederick, Maryland, that has barely been touched since 1953. It looks like an exhibit in a down-at-heels museum of the Cold War. All that brightly colored Formica and the kitschy kitchen ornaments—breezy symbols of 1950s American optimism—haven't stood the test of time.

Eric Olson's house—and Eric would be the first to admit this—could do with some redecoration.

Eric was born there, but he never liked Frederick and he never liked the house. He got out as quickly as he could after high school and ended up in Ohio and India and New York and Massachusetts, back in Frederick and Stockholm and California, but in 1993 he thought he would just crash out for a few months, and then ten years passed, during which time he hasn't decorated for three reasons:

1. He hasn't any money.

2. His mind is on other things.

3. And, really, his life ground to a shuddering halt on November 28, 1953, and if your living environment is

205

meant to reflect your inner life, Eric's house does the job. It is an inescapable reminder of the moment Eric's life froze. Eric says that if he ever forgets "why I'm doing this," he just needs to look around his house, and 1953 comes flooding back to him.

Eric says 1953 was probably the most significant year in modern history. He says we're all stuck in 1953, in a sense, because the events of that year have a continual and overwhelming impact on our lives. He rattled through a list of key events that occurred in 1953. Everest was conquered. James Watson and Francis Crick published, in *Nature* magazine, their famous paper mapping the double helix structure of DNA. Elvis first visited a recording studio, and Bill Haley's "Rock Around the Clock" gave the world rock and roll, and, subsequently, the teenager. President Truman announced that the United States had developed a hydrogen bomb. The polio vaccine was created, as was the color TV. And Allen Dulles, the director of the CIA, gave a talk to his Princeton alumni group in which he said, "Mind warfare is the great battlefield of the Cold War, and we have to do whatever it takes to win this."

On the night of November 28, 1953, Eric went to bed, as normal, a happy nine-year-old child. The family home had been built three years earlier, and his father, Frank, was still putting the finishing touches to it, but now he was in New York on business. Eric's mother, Alice, was sleeping down the hall. His little brother, Nils, and his sister, Lisa, were in the next room.

And then, somewhere around dawn, Eric was woken up.

"It was a very dim November predawn," Eric said.

Eric was woken up by his mother and taken down the hall, still wearing his pajamas, toward the living room—the same room where the two of us now sat, on the same sofas.

Eric turned the corner to see the family doctor sitting there.

"And," Eric said, "also, there were these two . . ." Eric searched for a moment for the right word to describe the others. He said, "There were these two . . . men . . . there also."

The news that the men delivered was that Eric's father was dead.

"What are you *talking* about?" Eric asked them, crossly.

"He had an accident," said one of the men, "and the accident was that he fell or jumped out of a window."

"Excuse me?" said Eric. "He did *what*?"

"He fell or jumped out of a window in New York."

"What does that *look* like?" asked Eric.

This question was greeted with silence. Eric looked over at his mother and saw that she was frozen and empty-eyed.

"How do you fall out of a window?" said Eric. "What does that mean? Why would he do that? What do you mean: fell or jumped?"

"We don't know if he fell," said one of the men. "He might have fallen. He might have jumped."

"Did he *dive*?" asked Eric.

"Anyhow," said one of the men, "it was an accident."

"Was he standing on a *ledge* and he jumped?" asked Eric.

"It was a work-related accident," said one of the men.

"Excuse me?" said Eric. "He fell out a window and that's *work* related? What?"

Eric turned to his mother.

"Um," he said. "What is his work again?"

Eric believed his father was a civilian scientist, working with chemicals at the nearby Fort Detrick military base.

Eric said to me, "It very quickly became an incredibly rancorous issue in the family because I was always the kid saying, 'Excuse me, where did he *go*? Tell me this story again.' And my mother very quickly adopted the stance, 'Look, I've told you this story a *thousand* times.' And I would say, 'Yeah, but I don't get it.'"

Eric's mother had created—from the same scant facts offered to Eric—this scenario: Frank Olson was in New York. He was staying on the tenth floor of the Statler Hotel, now the Pennsylvania Hotel, across the road from Madison Square Garden, in midtown Manhattan. He had a bad dream. He woke up confused, and headed in the dark toward the bathroom. He became disorientated and fell out of the window.

It was 2:00 A.M.

Eric and his little brother, Nils, told their school friends that their father had died of a "fatal nervous breakdown" although they had no idea what that meant.

Fort Detrick was what glued the town together. All their friends' fathers worked at the base. The Olsons still got invited to neighborhood picnics and other community events, but there didn't seem any reason for them to be there anymore.

When Eric was sixteen, he and Nils, then twelve, decided to cycle from the end of their driveway to San Francisco. Even at that young age, Eric saw the 2,415-mile journey as a metaphor. He wanted to immerse himself in unknown American terrain, the mysterious America that had, for some impenetrable reason, taken his father away from him. He and Nils would "reach the goal"—San Francisco—"by small continuous increments of motion along a single strand." This was in Eric's mind a test run for another goal he would one day reach in an equally fastidious way: the solution to the mystery of what happened to his father in that hotel room in New York at 2:00 A.M.

I spent a lot of time at Eric's house, reading his documents and looking though his photos and watching his home movies. There were pictures of the teenage Eric and his younger brother, Nils, standing by their bikes. Eric had captioned the photograph "Happy Bikers." There were 8 mm films shot in the 1940s and early 1950s of Eric's father, Frank, playing with the children in the garden. Then there were some films Frank Olson had shot himself during a trip he made to Europe a few months before he died. There was Big Ben and Changing the Guard. There was the Brandenberg Gate in Berlin. There was the Eiffel Tower. It looked like a family holiday, except the family wasn't with him. Sometimes, in these 8-mm films, you catch a glimpse of Frank's traveling companions, three men, wearing long dark coats and trilby hats, sitting in Parisian pavement cafés, watching the girls go by.

I watched them, and then I watched a home movie that a friend of Eric's had shot on June 2, 1994, the day Eric had his father's body exhumed.

There was the digger breaking through the soil.

There was a local journalist asking Eric, as the coffin was hauled noisily into the back of a truck, "Are you having second thoughts about this, Eric?"

She had to yell over the sound of the digger.

"Ha!" Eric replied.

"I keep expecting you to change your mind," shouted the journalist.

Then there was Frank Olson himself, shriveled and brown on a slab in a pathologist's lab at Georgetown University, Washington, his leg broken, a big hole in his skull.

And then, in this home video, Eric was back at home, exhilarated, talking on the phone to Nils: "I saw Daddy today!"

After Eric put the phone down he told his friend with the video camera the story of the bicycle trip he and Nils had taken in 1961, from the bottom of their driveway all the way to San Francisco.

"I'd seen an article in *Boys' Life* about a fourteen-year-old kid who cycled from Connecticut to the West Coast," Eric said, "so I figured my brother was twelve and I was sixteen and that averaged out at fourteen, so we could do it. We got these terrible heavy two-speed twin bikes, and we started off right here. Forty West. We heard it went all the way! And we made it! We went all the way!"

"No!" said Eric's friend.

"Yeah," said Eric. "We cycled across the country."

"No way!"

"It's an incredible story," said Eric. "And we've never heard of a younger person than my brother who cycled across the United States. It's doubtful there is one. When you think about it, twelve, and alone. It took us seven weeks, and we had unbelievable adventures all the way."

"Did you camp out?"

"We camped out. Farmers would invite us to stay in their houses. In Kansas City the police picked us up, figuring we were runaways, and when they found out we weren't they let us stay in their jail."

"And your mom let you do this?"

"Yeah, that's a kind of unbelievable mystery."

(Eric's mother, Alice, had died by 1994. She had been drinking on the quiet since the 1960s, and had begun locking herself in the bathroom and coming out mean and confused. Eric would never have exhumed his father's remains while she was alive. His sister, Lisa, had died too, together with her husband and their two-year-old son. They'd been flying to the Adirondacks, where they were going to invest money in a lumber mill. The plane crashed, and everyone on board was killed.)

"Yeah," said Eric, "it's an unbelievable mystery that my mother let us go, but we called home twice a week from different places, and the local paper, the Frederick paper, twice a week had these front-page articles like *Olsons Reach St. Louis!* All across the country back then there were billboards advertising a place called Harold's Club, which was a big gambling casino in Reno. It used to be the biggest casino in the world. And their motif was HAROLD'S CLUB OR BUST! Every day we'd see these billboards: HAROLD'S CLUB OR BUST! It became a kind of slogan for our journey. When we got to Reno we realized we couldn't get into Harold's Club because we were too young. So we decided to make a sign that said HAROLD'S CLUB OR BUST! and tie it to the back of our bikes, go over to Harold's Club, and tell Harold, whoever he might be, that we'd carried this across the whole United States and we were just crazy to see Harold's Club. So we went into a drugstore. We got an old cardboard box and bought some crayons, and we started writing this sign. The woman who sold us the crayons said, "What are you guys doing?"

"We said, 'We're going to make a sign, HAROLD'S CLUB OR BUST! and tell Harold that we cycled all the way from . . .'"

"She said, 'These people are very smart. They're not going to fall for this.'

"So we made this thing, took it out onto the streets, scuffed it up, tied it to the back of our bikes, went over to Harold's Club, got to this big entryway—Harold's Club was this gigantic thing, literally the biggest gambling casino in the world—and there was a doorperson there.

"He said, 'What do you boys want?'

"We said, 'We want to meet Harold.'

"He said, 'Harold is not here.'

"We said, 'Well, who *is* here?'

"He said, 'Harold senior is not here but Harold junior is here.'

"We said, 'That's fine, we'll take Harold junior.'

"He said, 'Okay, I'll go in and see.'

"Pretty soon out strides this dude in a fancy cowboy suit. Handsome guy. So he comes out and looks at our bikes and he says, 'What are you guys doing?'

"We said, 'Harold. We've been cycling across the United States and we've wanted to see Harold's Club the whole time. We've been sweating across the desert.'

"And he said, 'Well, *come on in*!'

"We ended up staying for a week at Harold's Club. He took us up in a helicopter around Reno, put us up in a fancy hotel.

And when we were leaving he said, 'I guess you guys want to see Disneyland, right? Well, let me call up my friend Walt!'

"So he called up Walt Disney—and this is one of the great disappointments of my life—Walt wasn't home."

I have wondered why Eric spent the evening of the day he had his father's body exhumed telling his friend the story of Harold's Club or Bust. Maybe it's because Eric had spent so much of his adult life failing to be offered the kindness of strangers, failing to benefit from anything approaching an

Happy Bikers.

American dream, but now Frank Olson was out there, lying on a slab in a pathologist's lab, and perhaps things were about to turn around for Eric. Maybe some mysterious Harold Junior would come along and kindly explain everything.

In 1970, Eric enrolled at Harvard. He went home every Thanksgiving weekend, and because Frank Olson went out of the window during the Thanksgiving holiday of 1953, the family invariably ended up watching old home movies of Frank, and Eric inevitably said to his mother, "Tell me the story again."

During Thanksgiving 1974, Eric's mother replied, "I've told you this story a hundred times, a thousand times."

Eric said, "Just tell me it one more time."

And so Eric's mother sighed and she began.

Frank Olson had spent a weekend on an office retreat in a cabin called the Deep Creek Lodge in rural Maryland. When he came home, his mood was unusually anxious.

He told his wife, "I made a terrible mistake and I'll tell you what it was when the children have gone to bed."

But the conversation never got around to what the terrible mistake had been.

Frank remained agitated. He told Alice he wanted to quit his job and become a dentist. On Sunday night Alice tried to calm him down by taking him to the cinema in Frederick to see whatever was on, which turned out to be a new film titled *Martin Luther*.

It was the story of Luther's crisis of conscience over the corruption of the Catholic Church in the sixteenth century, when its theologians claimed it was impossible for the

Church to do any wrong, because they defined the moral code. They were fighting the Devil, after all. The film climaxed with Luther declaring, "No. Here I stand, I can do no other." The moral of *Martin Luther* is that the individual cannot hide behind the institution.

(*TV Guide*'s movie-review database gives *Martin Luther* two out of five stars and says, "It is not 'entertainment' in the usual sense of the word. One wishes there might have been some humor in the script, to make the man look more human. The film was made with such respect that the subject matter seems gloomy when it should be uplifting.")

The trip to the cinema didn't help Frank's mood, and the next day it was suggested by some colleagues that he go to New York to visit a psychiatrist. Alice drove Frank to Washington, D.C., and dropped him off at the offices of the men who would accompany him to New York.

This was the last time she ever saw her husband.

On the spur of the moment, during that Thanksgiving weekend in 1974, Eric asked his mother a question he'd never thought to ask before: "Describe the offices where you dropped him off."

So she did.

"Jesus Christ," said Eric, "that sounds just like CIA headquarters."

And then Eric's mother became hysterical.

She screamed, "You will *never* find out what happened in that hotel room!"

Eric said, "As soon as I finish at Harvard I'm going to move back home and I'm not going to rest until I find out the truth."

Eric didn't have to wait long for a breakthrough. He received a telephone call from a family friend on the morning of June 11, 1975: "Have you seen *The Washington Post*? I think you'd better take a look."

It was a front-page story, and the headline read:

SUICIDE REVEALED

A civilian employee of the Department of the Army unwittingly took LSD as part of a Central Intelligence Agency test, then jumped 10 floors to his death less than a week later, according to the Rockefeller Commission report released yesterday.*

The man was given the drug while attending a meeting with CIA personnel working on a test project that involved the administration of mind-bending drugs to unsuspecting Americans.

"This individual was not made aware he had been given LSD until about 20 minutes after it had been administered," the commission said. "He developed serious side effects and was sent to New York with a CIA escort for psychiatric treatment. Several days later, he jumped from a tenth-floor window of his room and died as a result."

The practice of giving drugs to unsuspecting people lasted from 1953 to 1963, when it was discovered by the CIA's inspector general and stopped, the commission said.

---

* The Rockefeller Commission had been created to investigate CIA misdeeds in the aftermath of the Watergate scandal.

Is this my father? thought Eric.

The headline was misleading. Not much was "revealed"— not even the name of the victim.

Is this what happened at the Deep Creek Lodge? thought Eric. *They slipped him LSD? No, but it has to be my father. How many army scientists were jumping out of hotel windows in New York in 1953?*

On the whole, the American public reacted to the Frank Olson story in much the same way as they responded, fifty years later, to the news that Barney was being used to torture Iraqi detainees. Horror would be the wrong word. People were basically amused and fascinated. As in the case of Barney, this response was, I think, triggered by the disconcertingly surreal combination of dark intelligence secrets and familiar pop culture.

"For America it was lurid," said Eric, "and exciting."

The Olsons were invited to the White House so that President Ford could personally apologize to them—"He was very, *very* sorry," said Eric—and the photographs from that day show the family beaming and entranced inside the Oval Office.

"When you look at those photographs now," I asked Eric one day, "what do they say to you?"

"They say that the power of that Oval Office for seduction is enormous," Eric replied, "as we now know from Clinton. You go into that sacred space—that oval—and you're really in a special charmed circle and you can't think straight. It works. It really works."

Outside the White House, after their seventeen-minute meeting with President Ford, Alice Olson gave a statement to the press.

"I think it should be noted," she said, "that an American family can receive communication from the President of the United States. I think that's a tremendous tribute to our country."

"She felt very embraced by Gerald Ford," said Eric. "They laughed together, and so on."

The president promised the Olsons full disclosure, and the CIA provided the family, and America, with a flurry of details, each more unexpected than the last.

The CIA had slipped LSD into Frank Olson's Cointreau at a camping retreat called the Deep Creek Lodge. The project was codenamed MK-ULTRA, and they did it, they explained, because they wanted to watch how a scientist would cope with

The Olsons in the Oval Office. Eric is second from the right.

The Olsons back at home.

the effects of a mind-altering drug. Would he be unable to resist revealing secrets? Would the information be coherent? Could LSD be used as a truth serum for CIA interrogators?

And there was another motive. The CIA later admitted that they very much enjoyed paranoid thrillers like *The Manchurian Candidate,* and they wanted to know if they could create real-life brainwashed assassins by pumping people with LSD. But Frank Olson had had a bad trip, perhaps giving rise to the legend that if you take LSD you believe you can fly and you end up falling out of windows.

Social historians and political satirists immediately labeled these events "a great historical irony," and Eric repeated these words to me through gritted teeth because he doesn't appreciate the fact that his father's death has become a fragment of an irony.

"The great historical irony," Eric said, "being that the CIA brought LSD to America, thereby bringing a kind of enlightenment, thereby opening up a new level of political consciousness, thereby sowing the seeds of its own undoing because it created an enlightened public. It made great copy, and you'll find that this theme is the motif of a lot of books."

The details kept coming, so thick and fast that Frank Olson was in danger of becoming lost, swept away like a twig in the tidal wave of this colorful story. The CIA also told the Olsons that in 1953 they created an MK-ULTRA brothel in New York City, where they spiked the customers' drinks with LSD. They placed an agent named George White behind a one-way mirror where he molded, and passed up the chain of command, little models made out of pipe cleaners. The models represented the sexual positions considered, by the observant George White, to be the most effective in releasing a flow of information.

When George White left the CIA, his letter of resignation read, in part: "I toiled wholeheartedly in the vineyards because it was fun, fun, fun. . . . Where else could a red-blooded American boy lie, kill, cheat, steal, rape and pillage with the sanction and blessing of the all-highest?"

George White addressed this letter to his boss, the very same CIA man who had spiked Frank Olson's Cointreau: an ecology-obsessed Buddhist named Sidney Gottlieb.

Gottlieb had learned the art of sleight of hand from a Broadway magician named John Mulholland. This magician is all but forgotten today but back then he was a big star, a David Copperfield, who mysteriously bowed out of the public eye in 1953, claiming ill health, when the truth was that he had been secretly employed by Sidney Gottlieb to teach agents how to spike people's drinks with LSD. Mulholland also taught Gottlieb how to slip biotoxins into the tooth-brushes and cigars of America's enemies abroad.

It was Gottlieb who traveled to the Congo to assassinate the country's first democratically elected prime minister, Patrice Lumumba, by putting toxins in his toothbrush (he failed: the story goes that someone else, a non-American, managed to assassinate Lumumba first). It was Gottlieb who mailed a monogrammed handkerchief, doctored with brucel-losis, to Iraqi colonel Abd al-Karim Qasim. Qasim survived. And it was Gottlieb who traveled to Cuba to slip poisons into Fidel Castro's cigars and his diving suit. Castro survived.

It was like a comedy routine—the Marx Brothers Become Covert Assassins—and sometimes it seemed to Eric as if his family were the only people not laughing.

"The image that was presented to us," Eric said, "was fra-ternity boys out of control: 'We tried some crazy things, and we made errors of judgment. We put various poisons in Castro's cigars but none of that worked. And then we decid-ed that we weren't really good at that sort of thing.'"

"A clown assassin," I said.

"A clown assassin," said Eric. "Ineptitude. We drug peo-ple and they jump out of windows. We try to assassinate people and we get there too late. And we never actually assassinated anybody."

Eric paused.

"And Gottlieb turns up everywhere!" he said. "Is Gottlieb the only person in the shop? Does he have to do *every-thing*?" Eric laughed. "And this is what my mother was seizing on when she talked to Gottlieb. She said, 'How could you do such a harebrained scientific experiment? Where's the medical supervision? Where's the control group? You call this *science*?' And Gottlieb basically replied, 'Yeah, it was a bit casual. We're sorry for that.'"

As I sat in Eric Olson's house and listened to his story I remembered that I had heard Sidney Gottlieb's name mentioned before, in some other faraway context. Then it came to me. Before General Stubblebine came along, the secret psychic spies had another administrator: Sidney Gottlieb.

It took me a while to remember this because it seemed so unlikely. What was someone like Sidney Gottlieb, a poisoner, an assassin (albeit a not particularly good one), the man indirectly responsible for the death of Frank Olson, doing in the middle of this other, funny, psychic story? It seemed remarkable to me that the organizational gap in the intelligence world between the light side (psychic supermen) and the dark side (covert assassinations) has been so narrow. But it wasn't until Eric showed me a letter his mother received out of the blue on July 13, 1975, that I began to understand just how narrow it was. The letter was from the Diplomat Motor Hotel, in Ocean City, Maryland. It read:

Dear Mrs. Olson,

   After reading the newspaper accounts on the tragic death of your husband, I felt compelled to write to you.

   At the time of your husband's death, I was the assistant night manager at the Hotel Statler in New York and was at his side almost immediately after his fall. He attempted to speak but his words were unintelligible. A priest was summoned and he was given the last rites.

Having been in the hotel business for the last 36 years and wit-nessed innumerable unfortunate incidents, your husband's death disturbed me greatly, due to the most unusual circumstances of which you are now aware.

If I can be of any assistance to you, please do not hesitate to call upon me.

My heartfelt sympathy to you and your family.

Sincerely,

Armond D. Pastore

General Manager

The Olsons did phone Armond Pastore to thank him for his letter, and it was then that Pastore told them what had happened in the moments after Frank died in his arms on the street at 2:00 A.M.

Pastore said he went back inside the hotel and spoke to the telephone switchboard operator. He asked her if any calls had been made from Frank Olson's room.

She said that there was just one call, and she had listened in to it. It was very short. It was made immediately after Frank Olson went out the window.

The man in Frank Olson's room said, "Well, he's gone."

The voice on the other end of the phone said, "That's too bad."

And then they both hung up.

15. HAROLD'S CLUB OR BUST!

Eric Olson has a swimming pool in his back garden—one of the very few additions to the house made since 1953. On a hot day in August, Eric; his brother, Nils; Eric's son, who usually lives in Sweden; Nils's wife and their children; and some of Eric's friends and I were sunbathing by the pool, when a truck covered with pictures of party balloons—Capital Party Rentals—pulled up in his driveway to drop off one hundred plastic seats.

"Hey! Colored chairs!" yelled Eric.

"You want the colored chairs?" said the driver.

"Nah," said Eric. "Inappropriate. I'll take the gray ones."

Eric had brought a ghetto blaster down to the poolside and he tuned it to National Public Radio's *All Things Considered* because the legendary reporter Daniel Schorr was going to deliver a commentary about him. Daniel Schorr was the first man to interview Khrushchev, he won three Emmys for his coverage of Watergate, and now he was turning his attention to Eric.

His commentary began.

. . . Eric Olson is ready to charge, in a news conference tomorrow, that the story of a suicide plunge makes no sense . . .

Eric leaned up against the wire fence that surrounded his swimming pool and grinned at his friends and family, who were listening intently to this broadcast.

. . . And that his father was killed to silence him about the lethal activities he'd been involved in, projects codenamed Artichoke and MK-ULTRA. Today, a spokesman for the CIA said no congressional or executive branch probes of the Olson case have turned up any evidence of homicide. Eric Olson may not have the whole story. The thing is, the government's lid on its secrets remains so tight, we may never know the whole story. . . .

Eric flinched.

"Don't *go* there, Dan," he muttered to himself. "Don't go there."

. . . This is Daniel Schorr . . .

"Don't *go* there, Dan," Eric said.

He turned to us all, sitting by the pool. We sat there and said nothing.

"See?" said Eric. "That's what they want to do. 'We may never know the whole story.' And there's so much comfort they take in that. Bullshit. *Bullshit.* 'Oh, it could be this, it could be that, and everything in the CIA is a hall of mir-

rors . . . layers . . . you can never get to the bottom . . . '
When people say that, what they're really saying is, 'We're
comfortable with this because we don't *want* to know.' It's
like my mother always said, 'You're never going to know
what happened in that hotel room.' Well, something *did*
happen in that room and it is *knowable.*"

Suddenly, Eric is sixty years old. Decades have gone by,
and he has spent them investigating his father's death. One
day I asked him if he regretted this, and he replied, "I regret
it *all* the time."

Piecing together the facts has been hard enough for Eric, the
facts being buried in classified documents, or declassified documents covered with thick black lines made with marker pens,
or worse. Sidney Gottlieb admitted to Eric during one meeting
that he had, on his retirement, destroyed the MK-ULTRA files.
When Eric asked him why, Gottlieb explained that his "ecological sensitivity" had made him aware of the dangers of
"paper overflow."

Gottlieb added that it didn't really matter that the documents were ruined, because it was all a waste anyway. All the
MK-ULTRA experiments were futile, he told Eric. They had
all come to nothing. Eric left Gottlieb realizing that he'd
been beaten by a truly first-class mind.

What a brilliant cover story, he thought. *In a success-
obsessed society like this one, what's the best rock to hide
something under? It's the rock called failure.*

So, most of the facts were retained only in the memories of
men who did not want to talk. Nonetheless, Eric has constructed a narrative that is just as plausible as, even more
plausible than, the LSD suicide story.

Collecting the facts has been difficult enough, but there has been something even harder.

"The old story is so much fun," Eric said, "why would anyone want to replace it with a story that's *not* fun. You see? The person who puts the spin on the story controls it from the beginning. It's very hard for people to read against the grain of what you've been told the narrative is about."

"Your new story is not as much fun," I agreed.

"This is no longer a happy, feel-good story," Eric said, "and I don't like it any better than anyone else does. It's hard to accept that your father didn't die because of suicide, nor did he die because of negligence after a drug experiment, he died because they *killed* him. That's a different feeling."

And, vexingly for Eric, on the rare occasions when he's convinced a journalist that the CIA *murdered* his father, the revelation has not been greeted with horror. One writer declined Eric's invitation to attend his press conference, saying, "We *know* the CIA kills people. That's old news."

In fact, Eric told me, this would be the first time anyone had ever publicly charged the CIA with murdering an American citizen.

"People have been so brainwashed by *fiction,*" said Eric as we drove to the local Kinko's to pick up the press releases for the conference, "so brainwashed by the Tom Clancy thing, they think, 'We *know* this stuff. We *know* the CIA does this.' Actually, we know *nothing* of this. There's *no* case of this, and all this fictional stuff is like an immunization against reality. It makes people think they know things that they don't know and it enables them to have a kind of superficial quasi-sophistication and cynicism which is just a thin layer beyond which they're not cynical at all."

It isn't that people aren't interested: it's that they're interested in the wrong way. Recently, a theater director approached Eric for his permission to turn the Frank Olson story into "an opera about defenestration," but Eric declined, explaining that this was a complex enough tale anyway even without having the facts *sung* at an audience. Tomorrow's press conference was really Eric's last chance to convince the world that his father was not an LSD suicide.

There were so many ways for Eric to recount his new version of the story at the press conference. It was impossible for him—for anyone—to know how to do it in the most coherent and still entertaining way. Eric's new story is not only no longer fun, it is exasperatingly intricate. There's so much information to absorb that an audience could just glaze over.

Really, this story begins with the proclamation delivered by the CIA director Allen Dulles to his Princeton alumni group in 1953.

"Mind warfare," he said, "is the great battlefield of the Cold War, and we have to do whatever it takes to win this."

Before Jim Channon and General Stubblebine and Colonel Alexander came along, there was Allen Dulles, the first great out-of-the-box thinker in U.S. intelligence. He was a great friend of the Bushes, and was once the Bush family lawyer, a pipe-smoking patriarch who believed that the CIA should be like an Ivy League university, taking inspiration not only from agents but from scientists, academics, and whoever else might come up with something new. It was Dulles who moved the CIA's headquarters from central Washington, D.C., to suburban Langley, Virginia (now renamed the George Bush Center for Intelligence), because he wanted to create a thoughtful, out-of-town campus milieu. It was Dulles who sent undercover CIA agents out into the American suburbs in the 1950s and 1960s to infiltrate séances in the hope of unearthing and recruiting America's most talented clairvoyants to his mind-warfare battlefield, which is how the relationship between intelligence and the psychic world was born. But it was General Stubblebine who, inspired by the First Earth Battalion, proclaimed a generation later that anyone could be a great psychic, and so opened the doors wide, and Major Ed Dames joined the program, and subsequently revealed the secrets of the unit on the Art Bell show, and then all hell broke loose and, through no fault of any of the military personnel involved, thirty-nine people in San Diego killed themselves in an attempt to hitch a ride on Prudence and Courtney's Hale-Bopp companion.

Allen Dulles put Sidney Gottlieb in charge of the fledgling psychic program, and also of MK-ULTRA, and then a third covert mind-warfare project known as Artichoke.

Artichoke is the program that is not fun.

Recently declassified documents reveal that Artichoke was all about inventing insane, brutal, violent, frequently fatal new ways of interrogating people.

Frank Olson was not just a civilian scientist working with chemicals at Fort Detrick. He was a CIA man too. He was working for Artichoke. That is why he was in Europe in the months before he died, sitting in sidewalk cafés with the men wearing long coats and trilbies. They were there on Artichoke business. Eric's father was—and there is no pleasant way of putting this—a pioneering torturer, or at the very least a pioneering torturer's assistant. Artichoke was the First Earth Battalion of torture—a like-minded group of groundbreaking out-of-the-box thinkers, coming up with all manner of clever new ways of getting information out of people.

An example: According to a CIA document dated April 26, 1952, the Artichoke men "used heroin on a routine basis" because they determined that heroin (and other substances) "can be useful in reverse because of the stresses produced when they are withdrawn from those who are addicted to their use."

This is why, Eric has learned, his father was recruited to Artichoke. He, alone among the interrogators, had a scientific knowledge of how to administer drugs and chemicals.

And now, in 2004, this Artichoke-created cold-turkey method of interrogation is back in business. Mark Bowden, the author of *Black Hawk Down*, interviewed a number of CIA interrogators for the October 2003 edition of *Atlantic Monthly*, and this is the scenario he constructed:

> On what may or may not have been March 1 [2003] the notorious terrorist Khalid Sheikh Mohammed was roughly awakened by a raiding party of Pakistani and American commandos. . . . Here was the biggest catch yet in the war on terror. Sheikh Mohammed is considered the architect of

two attempts on the World Trade Center: the one that failed, in 1993, and the one that succeeded so catastrophically, eight years later. . . . He was flown to an "undisclosed location" (a place the CIA calls "Hotel California")—presumably a facility in another cooperative nation, or perhaps a specially designed prison aboard an aircraft carrier.

It doesn't much matter where, because the place would not have been familiar or identifiable to him. Place and time, the anchors of sanity, were about to come unmoored. He might as well have been entering a new dimension, a strange new world where his every word, move, and sensation would be monitored and measured; where things might be as they seemed but might not; where there would be no such thing as day or night, or normal patterns of eating and drinking, wakefulness and sleep; where hot and cold, wet and dry, clean and dirty, truth and lies, would all be tangled and distorted.

The space would be filled night and day with harsh light and noise. Questioning would be intense—sometimes loud and rough, sometimes quiet and friendly, with no apparent reason for either. The session might last for days, with interrogators taking turns, or it might last only a few minutes. On occasion he might be given a drug to elevate his mood prior to interrogation; marijuana, heroin, and sodium pentothal have been shown to overcome a reluctance to speak. These drugs could be administered surreptitiously with food or drink, and given the bleakness of his existence, they might even offer a brief period of relief and pleasure, thereby creating a whole new category of longing—and new leverage for his interrogators.

See how in this scenario a slice of Jim Channon's First Earth Battalion ("harsh light and noise") and a slice of Frank Olson's Artichoke ("a whole new category of longing") come together like two pieces of a jigsaw.

On the day before Eric's press conference, Eric and I watched old 8-mm home movies of his father playing in the garden with his children. On the screen, Frank was riding a wobbly old bicycle and Eric, then a toddler, was resting on the handlebars. Eric gazed, smiling, at the screen.

He said, "*There's* my father. Right there! That's him! In comparison with the other guys from the CIA, he has an open face. Um . . ." Eric paused. "Basically," he said, "this is a story about a guy who had a simple moral code and a naive view of the world. He wasn't fundamentally a military guy. And he certainly wasn't someone who would be involved in 'terminal interrogations.' He went though a moral crisis, but he was in too deep and they couldn't let him out."

We continued watching the home video. Then Eric said, "Think of how much could have been different if he was alive to tell any of this. Ha! The whole history of a lot of things would be different. And you can see a lot of that just in his face. A lot of the other men have very tight, closed faces. He doesn't . . ." And Eric trailed off.

At some point during his investigation, Eric hooked up with the British journalist Gordon Thomas, who has written numerous books on intelligence matters. Through Thomas, Eric learned that during a trip to London in the summer of 1953 his father had apparently confided in William Sargant, a consultant psychiatrist who advised British intelligence on brainwashing techniques.

According to Thomas, Frank Olson told Sargant that he had visited secret joint American-British research installations near Frankfurt, where the CIA was testing truth serums on "expendables," captured Russian agents and ex-Nazis. Olson confessed to Sargant that he had witnessed something terrible, possibly "a terminal experiment" on one or more of the expendables. Sargant heard Olson out and then reported to British intelligence that the young American scientist's misgivings were making him a security risk. He recommended that Olson be denied further access to Porton Down, the British chemical-weapons research establishment.

After Eric learned this, he told his friend, the writer Michael Ignatieff, who published an article about Eric in *The New York Times*. A week later, Eric received the telephone call he'd been waiting for his whole life. It was a real Harold Junior, one of his father's best friends from Fort Detrick, a man who knew everything, and was willing to tell Eric the whole story.

His name was Norman Cournoyer.

Eric spent a weekend at Norman's house in Connecticut. Revealing to Eric the secrets he'd been harboring all these years was so stressful for Norman that he repeatedly excused himself so he could go to the toilet to vomit.

Norman told Eric that the Artichoke story was true. Frank had told Norman that "they didn't mind if people came out of this or not. They might survive, they might not. They might be put to death."

Eric said, "Norman declined to go into detail about what this meant but he said it wasn't nice. Extreme torture, extreme use of drugs, extreme stress."

Norman told Eric that his father was in deep and horrified at the way his life had turned. He watched people die in Europe, perhaps he even helped them die, and by the time he returned to America he was determined to reveal what he had seen. There was a twenty-four-hour contingent of Quakers down at the Fort Detrick gates, peace protesters, and Frank would wander over to chat with them, much to the dismay of his colleagues. Frank asked Norman one day, "Do you know a good journalist I can talk to?"

And so, Eric said, slipping LSD into his father's Cointreau at the Deep Creek Lodge was not an experiment that went wrong: it was designed to get him to talk while hallucinating. And Frank failed the test. He revealed his intentions to Gottlieb and the other MK-ULTRA men present. This was the "terrible mistake" he had made. Seeing *Martin Luther* on the Sunday night had made him all the more determined to quit his job. *Here I stand. I can do no other.*

And on the Monday morning Frank did, indeed, tender his resignation, but his colleagues persuaded him to seek psychological counseling in New York.

Documents reveal that Frank never saw a psychiatrist in New York. He was taken instead, by Gottlieb's deputy, to the office of the former Broadway magician John Mulholland, who probably hypnotized him, and Frank probably failed that test too.

Housing a possibly deranged and desperate man in a hotel room high above Seventh Avenue no longer seemed a regrettable error of judgment. It seemed the prelude to murder.

When Eric had his father's body exhumed in 1994, the pathologist, Dr. James Starrs, found a hole in Frank's head that—he concluded—came from the butt of a gun and not a fall from a tenth-floor window.

There were around forty journalists at Eric's press conference—crews from all the networks and many of the big newspapers. Eric had decided—for the purposes of clarity—to tell the story primarily through the narrative of his weekend with Norman Cournoyer. He repeatedly stressed that this was no longer a family story. This was now a story about what happened to America in the 1950s and how that informs what is happening today.

"Where's the *proof*?" asked Julia Robb, the reporter from Eric's local paper, the *Frederick News Post,* when he had finished. "Does all this rest on the word of one man, your father's friend?"

Julia looked around her to make the point that this Norman Cournoyer wasn't even in attendance.

"No," said Eric. He looked exasperated. "As I've tried to tell you, it conceptually rests on the idea that there are two vectors in this story and they only intersect in one place."

There was a baffled silence.

"Are you in any way motivated by ideology over this?" the man from Fox News asked.

"Just a desire to know the truth." Eric sighed.

Later, as the journalists milled around, eating from the buffet laid out on picnic tables, the conversation among the Olsons and their friends turned to Julia Robb, the reporter from the *Frederick News Post*. Someone said he thought it was a shame that the most hostile journalist present represented Eric and Nils's local paper.

"Yeah, it is," said Nils. "It's painful to me. I'm a professional here in town. I have connections with local people as a dentist, and I see people on a daily basis who come in and read the local paper, and that affects me."

Nils looked over across the garden at Eric, who was saying something to Julia, but we couldn't hear what.

Nils said, "At times you go through a phase of believing that maybe the story *is* a bunch of hooey, and that it *was* just a simple LSD suicide and *that*"—Nils glanced at Julia—"can trigger a kind of shame spiral. It's like the feelings you've had in the middle of the night, at three A.M., when you're trying to get to sleep and you start having some *thought* and the thought spins you into another negative thought and it kind of spins out of control and you have to shake yourself and maybe turn the light on and get grounded in reality again."

Eric and Julia were arguing now. Julia said something to Eric and then she walked away, back to her car. (Later, Eric said to me that Julia seemed "incensed, as if the entire story made her furious in some deep way that she was completely at a loss to articulate.")

"I mean," said Nils, "America fundamentally wants to think of itself as being good, and that we're fundamentally right in what we're doing, and we have a very compelling responsibility for the free world. And looking at some of these issues is troubling, because if America does have a darker side it threatens your hold on your view of America and it's kind of like, 'Gee, if I pull out this one underpinning of the American consciousness, is this a house of cards? Does it really threaten the fundamental nature of America?'"

We drifted back down to the swimming pool, and an hour passed, and then Eric joined us. He'd been in the house on the telephone. He was laughing.

"You hear the latest?" he said.

"Bring me up-to-date," said Nils. "I'm dying to hear."

"Julia," said Eric, "called Norman. I just called her and she said, 'Eric, I'm glad you phoned. I just called Norman. He says he has no reason to believe that the CIA would murder Frank Olson.' I said, 'Julia, thanks for respecting my wishes about not calling Norman.' She said, 'Eric, I'm a *reporter*. I have to do what's necessary to get the story.'"

Eric laughed, although nobody else did.

And so I drove to Connecticut, to Norman Cournoyer's house. I was slightly shaken by the news of the telephone call between Julia Robb and Norman. Had I got Eric wrong? Was he some kind of fantasist?

Norman lives in a large white bungalow in an upmarket suburban street. His wife answered the door and led me into the living room, where Norman was waiting for me. He pointed to the table and said, "I dug out some old photographs for you."

They were of Norman and Frank Olson, arm in arm, somewhere in the middle of Fort Detrick, circa 1953.

"Did you tell the reporter from the *Frederick News Post* that you had no evidence to suggest that Frank was murdered by the CIA?" I asked.

"Yeah," said Norman.

"Why did you do that?" I asked.

"Over the phone?" said Norman. "I think a journalist is making a big mistake in trying to get somebody to talk over the phone."

"So you *do* think Frank was murdered?" I said.

"I'm sure of it," said Norman.

And then he told me something he hadn't told Eric.

"I saw Frank after he'd been given the LSD," he said. "We joked about it."

"What did he say?" I asked.

"He said, 'They're trying to find out what kind of guy I am. Whether I'm giving secrets away.'"

"You were joking about it?" I said.

"We joked about it because he didn't react to LSD."

"He wasn't tripping at all?" I said.

"Nah," said Norman. "He was laughing about it. He said, 'They're getting very, very uptight now because of what they believe I am capable of.' He really thought they were picking on him because he was the man who might give away the secrets."

"Was he going to talk to a journalist?" I asked.

"He came so close it wasn't even funny," said Norman.

"Did he come back from Europe looking very upset?" I asked.

"Yeah," said Norman. "We talked about a week, ten days, after he came back. I said, 'What happened to you, Frank? You seem awfully upset.' He said, 'Oh, you know . . . ' I must admit, in all honesty, it's just coming back to me now. He said . . ."

Suddenly, Norman fell silent.

"I don't want to go on further than that," he said. "There are certain things I don't want to talk about."

Norman looked out the window.

"It speaks for itself," he said.

Eric hoped his press conference would, at least, change the language of the reporting of the story. At best it would motivate some energetic journalist to take up the challenge and find an unequivocal smoking gun that proved Frank Olson was pushed out of the window.

But in the days that followed the press conference it became clear that every journalist had decided to report the story in much the same way.

Eric had finally found "closure."

He was on the way to being "healed."

He had "laid his mystery to rest."

He could "move on" now.

Perhaps we will "never know" what really happened to Frank Olson, but the important thing was that Eric had achieved "closure."

The story was fun again.

16. THE EXIT

June 27, 2004

Jim Channon faxes me his Iraq exit strategy. This is the same document he sent to the army's chief of staff, General Pete Schoomaker, after Donald Rumsfeld had asked the general to bring "creative" thinkers into the fold.

Jim's strategy begins:

> When we left Vietnam, we did so with our tails between our legs. We were leaving at an undignified pace. In the eyes of the world that watches, the last moments are as telling as the first.

THE FIRST EARTH BATTALION SOLUTION

1. A touching and heartfelt ceremony [consisting of] mothers, children, teachers, soldiers, nurses and doctors from both sides. Where possible children will carry the actual awards (i.e., medals, trophies, small statues) of apprecia-

tion and recognition to those [American and Iraqi sol-
diers] honored.

2. The ceremonial surroundings we design are themselves a
gift to the future of Iraq. We recommend that a beautiful
global village be built as a setting. It can showcase the
kinds of alternative energy, sanitation and agricultural
technology appropriate for this part of the world.

3. [The ceremony will include the giving of] gifts from other
parts of the world. United Nations translators will be
available to interpret those gifts. An elder might speak
and a teen might speak about the promise of cooperation.

June 29, 2004

Today, sovereignty is transferred from the coalition forces to
the new Iraqi government. Whoever organized the ceremony
obviously chose not to implement Jim's ideas:

> Behind silver miles of new razor wire, behind high concrete
> barriers stronger than most medieval fortifications, behind
> sandbags, five security checks, U.S. armoured vehicles, U.S.
> armoured soldiers, special forces of various countries and
> private security guards, an American bureaucrat handed a
> piece of paper to an Iraqi judge, jumped on a helicopter
> and left the country.
>
> The first thing reporters saw as they came into the sun-
> shine from the banal auditorium where the newly sworn-in
> Iraqi government hailed the new era was two U.S. Apache
> helicopter gunships, pirouetting low in the furnace sky.

Fear of the bombers gave the occasion all the pomp of an office leaving do. It lasted only 20 minutes.

—James Meek, *The Guardian*.

I suppose this has been a book about the changing relationship between Jim Channon's ideas and the army at large. Sometimes the army seems like a nation, and Jim a village somewhere in the middle, like Glastonbury, looked on fondly but basically ignored. At other times, Jim seems right in the heart of things.

Perhaps the story is this: In the late 1970s, Jim, traumatized from Vietnam, sought solace in the emerging human-potential movement in California. He took his ideas back into the army and they struck a chord with the top brass, who had never before seen themselves as new age, but in their post-Vietnam funk it all made sense to them.

But then, over the decades that followed, the army, being what it is, recovered its strength and saw that some of the ideas contained within Jim's manual could be used to shatter people rather than heal them. Those are the ideas that live on in the War on Terror.

The "bureaucrat" referred to in *The Guardian* article, Paul Bremer, may have left the country today, but he has left behind him in Iraq 160,000 troops, the vast majority of them American. According to a July 2004 report by the Institute for Policy Studies and Foreign Policy in Focus, 52 percent of those American soldiers are experiencing low morale, 15 percent have screened positive for traumatic stress, 7.3 percent for anxiety, and 6.9 percent for depression. The suicide rate among American soldiers has increased from an eight-year average of 11.9 per 100,000 to 15.6 per 100,000.

As of September 2004, a total of 1,175 coalition soldiers have been killed since the war began, including 1,040 Americans. Some 7,413 more have been wounded. Military hospitals have reported a sharp increase in the number of amputations—the result of an "improved" design of body armor that protects vital organs but not limbs.

Between 12,800 and 14,843 Iraqi civilians are now dead as a result of the U.S. invasion and the ensuing occupation, with 40,000 more injured. These figures are less exact because nobody has really been keeping count.

Eighty percent of Iraqis say they have "no confidence" in either the U.S. civilian authorities or the coalition troops, in part, I've no doubt, because of the photographs that detailed the methods of interrogation employed by military intelligence at Abu Ghraib.

I have had the strangest telephone call. It was from somebody I've written about in this book, a man who continues to work within the U.S. military. I almost didn't include what he told me because it is utterly outlandish and impossible to substantiate. But it also rings true. He said he'd tell me the secret on the condition that I didn't reveal his name.

Before I repeat what he said, I should explain why I believe it rings true.

First, outlandishness hasn't stopped them before.

I once asked Colonel Alexander if there had been some sort of post-9/11 renaissance of MK-ULTRA.

"Not necessarily LSD," I added, "but a nonlethal weapon type of MK-ULTRA. Take the Guantanamo Bay ghetto-blaster story. Surely the most likely explanation is that they were playing him some kind of mind-altering noise, buried somewhere below Fleetwood Mac."

"You're sounding ridiculous," he replied.

He was right. I was sounding just as ridiculous as I sounded when I asked friends of Michael Echanis if they knew whether Michael had ever been involved in "influencing live-stock from afar." But those were the cards this story had dealt me.

(Remember that the crazy people are not always to be found on the outside. Sometimes the crazy people are deeply embedded on the inside. Not even the most imaginative conspiracy theorist has ever thought to invent a scenario in which a crack team of Special Forces soldiers and major generals secretly try to walk through their walls and stare goats to death.)

"Listen," said Colonel Alexander, crossly. "Nobody who lived through the trauma of MK-ULTRA" (he was talking about the trauma on the intelligence side, the trauma of being found out, not the trauma on the Olsons' side) "would ever involve themselves in something like that again. Nobody who lived through all those congressional hearings, that media reaction . . ." He paused. Then he said, "Sure, you've got kids in intelligence who've read all about MK-ULTRA and think, 'Gee. That sounds cool. Why don't we try *that* out?' But you'd never get a reactivation at command level."

Of course, a bunch of young enthusiasts in military intelligence thinking "that sounds cool" is exactly how these things can spring to life, and have done before.

The other reason why I think the secret rings true revolves around the mystery of why Major Ed Dames decided one day to reveal on the Art Bell show the existence of the psychic spying unit. When I asked Major Dames in Maui what his motive for this was, he shrugged and a faraway look crossed his face and he said, "I didn't have any motive. I didn't have any motive at all."

But he said it in such a way as to lead me to think that he actually had a very shrewd, secret motive. At the time I put Ed's pointedly enigmatic half-smile down to his well-deserved reputation as a somewhat self-aggrandizing mystery monger.

Many people blamed Ed for the closure of the unit, and some smelled a conspiracy. Ed's former psychic colleague Lyn Buchanan once told me he'd come to believe there was *another* psychic unit, even more deeply hidden, and presumably with more glamorous offices than theirs, and that the reason why *their* unit was revealed to the world was to divert attention from this mysterious other psychic team. Lyn's implication was that Ed was instructed to reveal the secrets by some high-up cabal.

At the time, I didn't give this theory much credence. I have often found that people at the heart of perceived conspiracies are often conspiracy theorists themselves. (I remember once speaking with a high-ranking Freemason from their Washington, D.C., headquarters. He said to me, "Of course it is simply absurd to think that the Freemasons secretly rule the world, but I'll tell you who does secretly rule the world: the

Trilateral Commission.") I put Lyn's assertion down to that peculiar facet of the conspiracy world.

But now I'm not so sure.

After Lyn Buchanan had presented his theory to me I e-mailed Skip Atwater, the extremely levelheaded former psychic headhunter from Fort Meade. Skip had been deeply involved in the unit, in an administrative capacity, between 1977 and 1987. Was there, I asked him, any truth to what Lyn had said?

He e-mailed me back:

It is true that if asked about the use of remote viewing, psychics, or whatever, the CIA can now say something like, "There was a program, but it has since been closed." And that is a true statement, but it's not the whole truth. For security reasons, I cannot detail further information about non [Fort Meade] programs. I would suppose however, that since the years that I was privy to such information, these efforts have changed direction a bit and are now highly focused on counterterrorism. For reasons of security management, it would be customary to . . . well, perhaps I shouldn't go on at this point.

And that was the end of Skip's e-mail.

I know that almost every former psychic spy from the old Fort Meade unit received a telephone call from the intelligence services in the weeks that followed 9/11. They were told that if they had any psychic visions of future terrorist attacks they shouldn't hesitate to inform the authorities.

And they did, in their droves. Ed Dames had a terrible

vision of al-Qaeda sailing a boat full of explosives into a nuclear submarine in San Diego harbor.

"I knew bin Laden's people were clever," Ed said to me of his vision, "but I hadn't realized they were *that* clever."

Ed reported his findings to the California Coast Guard's office.

Uri Geller had his telephone call from Ron, but that is all I know about Uri and Ron.

A number of second-generation remote viewers (psychic spies who learned the trade from members of the Fort Meade unit and subsequently set up their own training schools) were also contacted by the intelligence communities post-9/11. One—a woman named Angela Thompson—had a vision of mushroom clouds over Denver, Seattle, and Florida. I was present at a reunion convention of the former military psychics at a Doubletree Hotel in Austin, Texas, in the spring of 2002, when Angela presented her mushroom-cloud findings. The conference room was full of retired psychic spies and intelligence officers. When Angela said "mushroom clouds over Denver, Seattle, and Florida," everybody in the room gasped.

Prudence Calabrese was in the room. All seemed to have been forgiven regarding the Heaven's Gate mass suicides, because the FBI telephoned Prudence in late September 2001 and asked her to let them know as soon as she had any visions of future terrorist attacks.

Prudence did indeed have a vision, she told me, a truly awful vision. She FedExed the details of her vision to the FBI. They thanked her and have been requesting more psychic information ever since, she said.

"What was the vision?" I asked her.

There was a short silence.

"Put it this way," she said. "London is an area of high concern. It's certainly an area we've looked at and there's reason to be concerned if you live in London."

"I live in London," I said.

Prudence tried to change the subject, but I wouldn't let her.

"When?" I asked.

"Two-thirty in the morning!" she snapped. Then she laughed and turned serious. "Really, we're not at liberty to give any more information on this."

"Is there *anything* else you can tell me?" I asked.

"We know enough to be certain that something is going to happen," she said, "and we know enough to know the general vicinity in which something will happen."

"A landmark?" I asked.

"Yeah," said Prudence.

"A Houses of Parliament–type landmark?" I asked.

"I'm not going to tell you," she said.

"Surely not Buckingham Palace," I said, shocked.

It was at this point that my interrogation of Prudence finally cracked her.

"It's London Zoo," she said.

London Zoo was about to be hit by a dirty bomb, she said, one so powerful it would knock over the nearby BT Tower.

"How do you *know* this?" I asked, visibly upset.

"The elephants," she said.

The elephants were screaming in agony in her psychic vision, Prudence explained. The pain of the London Zoo elephants was the most intense and powerful image she had received. Prudence had gathered a team of fourteen psychic employees, based in Carlsbad, near San Diego. All fourteen of them, she said, had felt the pain of the elephants.

When I returned home to the United Kingdom I discovered to my relief that the London Zoo elephants had, some months prior to Prudence's psychic vision, all been moved to the Whipsnade Wild Animal Park, in rural Bedfordshire, about thirty miles north of London. How could the elephants be collateral damage in a London Zoo dirty bomb if there weren't any elephants left at London Zoo?

I have wondered whether Tom Ridge's Department of Homeland Security has ever issued a nonspecific warning of a future terrorist attack based on intelligence provided by a psychic. I spent a few weeks trying to find out whether they had, but my calls got me nowhere and I gave up and the psychics drifted from my mind.

I hadn't spent much time thinking about the psychics until I received the telephone call out of the blue and the man on the other end said he had a secret to reveal as long as I promised to protect his identity.

"Okay," I said.

"Do you know about remote viewing?" he said.

"The psychic spies?" I asked.

"Yeah," he said. "There's a *lot* of interest in it again."

"I know *that*," I said.

I told him about Ed and Angela and Prudence and Uri and the mysterious Ron.

"I don't suppose you know who Ron is?" I asked him.

"I'm not talking about *those* remote viewers," he said. "They've got some new guys in, and they're using remote viewing in a *very* different way."

"Mmm?" I said.

"They're taking remote viewing out of the office," he said.

"I'm sorry?" I said.

"They're taking remote viewing *out—of—the—office.*"

"Okay, thanks," I said.

I had no idea what he was telling me but it didn't sound like a particularly good secret.

"Do you understand?" he said, exasperated. "Remote viewing is no longer *office based.*"

"Uh," I said.

I think he was beginning to suspect he had picked the wrong journalist to reveal his secret to.

"I'm sorry that I'm not savvy enough to understand what you are cryptically telling me," I said.

"What do you know about the history of remote viewing?" he asked, slowly.

"I know it was office based," I said.

"That's right," he said.

"And it is no longer so?" I said, my eyes narrowing.

"Oh, for fuck's sake," he said. "If it's no longer office based then . . ."

He paused. He had two choices. He could either continue to reveal the secret enigmatically—which was a method that was clearly annoying both of us a little—or he could just come right out and tell me. And so he did.

"Psychic assassins," he announced. "Cool, eh? They're

teaching the Special Op assassins, the Fort Bragg guys who go out into the field to track down and assassinate terrorists, how to be psychic. They used to rely on hard intelligence, but things are changing. Intelligence is so often flawed. So instead they're going back to the power of the mind."

"How does it work?" I asked.

"We drop a Special Op guy in a jungle or a desert or at a border," he said. "We know the target is a few miles away but we don't know exactly where. What do we do? Wait for the spy planes? Wait for an interrogator to crack a prisoner? Sure, we do these things, but now we can augment all that with the power of the mind."

"So the assassins," I said, "while waiting for hard intelligence, psychically envisage the location of their targets and start tracking straightaway?"

"Sure," he said. "The mind is making a very big comeback down at Fort Bragg."

July 15, 2004

I hear from Guy Savelli. He sounds excited and I assume there has finally been some movement on his looming al-Qaeda paranormal sting operation. The last time I spoke to Guy he was being deluged with calls from young martial arts enthusiasts based in axis-of-evil countries who wanted to learn how to kill goats just by staring at them. Ever since then Guy has been waiting for the go-ahead to start teaching the terrorists the stare while acting as a spy on behalf of the intelligence services, but it is yet to come.

I presume he is calling to tell me the latest news on this, but, he says, something else, something incredible, has happened. He has received a telephone call from Fort Bragg. Can he get down there "right away" to demonstrate his powers to a new commanding general who "sees the spiritual side"?

"I'm going this weekend," he says.

"Are you taking an animal with you?" I ask.

"Yeah," he says. "They want me to bring an animal."

"A goat?" I ask.

"My resources are limited," says Guy.

"A hamster?" I ask.

"All I can tell you," said Guy, "is that there will be some kind of animal involvement."

"Are we talking about a small animal, cheap to purchase?" I ask.

"Correct," confirms Guy.

"A hamster," I say.

A silence.

"Yes," says Guy. "I am taking a hamster out there and I'm going to blow their minds with it."

I hear Guy's wife say something to him on the other end of the phone.

"That's them on the other line now," says Guy, urgently. "I'll call you back."

"Guy!" I shout after him. "Ask them if I can come too!"

July 19, 2004

I haven't heard from Guy in four days. I e-mail him to ask if there has been any movement and he finally calls back.

"Everything seems to be coming together," he says.

"Have you been to Fort Bragg with a hamster yet?"

"It's more than that," says Guy. "They're trying to get what I'm doing classified. They're trying to get me into a deeper military position."

"What do you mean?" I ask.

"They want me to actually go with them some places. Some Middle Eastern places."

I ask Guy to tell me more, and he does.

After Jim Channon produced his *First Earth Battalion Operations Manual* his commanders invited him to create a real-life Warrior Monk unit, traveling the world with their supernatural powers. As I have explained earlier, Jim turned the offer down because he was rational enough to realize that walking through walls and so on were good ideas on paper, but weren't, necessarily, achievable skills in real life.

But now, Guy says, this is exactly what they want *him* to do. They want him to lead a Warrior Monk unit into Iraq.

"What sort of powers will you be equipped with?" I ask.

"Hopefully quite a few," says Guy, "because we'll have to go in without weapons."

"Why?" I ask.

"Because it's the peaceful and gentle way," says Guy. "These are good, kind men. They know they've been doing it all wrong in Iraq. Remember: the guys in the Abu Ghraib photographs *trained* at Fort Bragg. And they screwed up big.

254

They know this crap can't go on anymore. So now they've asked me to come down."

"And teach them how to stare people to death?" I ask.

"No," says Guy. "This is different now. This is such a revolutionary idea, it'll change the way they treat those prisoners. Think of what you can do just by staring. You can confuse people to the point where they don't know what the hell they're looking at, and they'll give you all sorts of information."

Guy says he hasn't yet told Special Forces that he's keeping me informed of all developments.

"Won't they be furious?" I ask him.

"Nah," says Guy. "This is the kind and the gentle way. They'll *want* people to know about this."

"Next time you go to Fort Bragg with a hamster," I say, "can I come?"

"I'll ask them," says Guy, "when the time is right."

July 23, 2004

Guy calls. He has been to Fort Bragg with a hamster.

"I'll tell you, Jon," he says. "The Special Forces guys came into the meeting in a pretty hostile frame of mind and they left like little kids. They're frustrated. They're afraid. They know they're senselessly screwing up in Iraq. And they know that their only alternative is *me*. Thought projection is really sinking in with these guys. They definitely, one hundred percent, want to go back to the old ways."

"So you're going to Iraq?" I ask.

"Looks that way," says Guy.

"When?" I ask.

"We have a limited time before we leave," says Guy.

"Have you told them yet that you're telling me everything?" I ask.

"Nah," says Guy. "But they'll be fine with it. I'm sure you'll be able to come with me next time. It'll be great PR for them. And there's another reason why I know they'll want you on board. If the enemy knows we have this *power* it'll scare the shit out of them."

Guy pauses.

"I'm going to tell them all about you tomorrow," he says.

July 28, 2004

I phone Guy Savelli repeatedly today, as I have done all week, but still to no avail. He doesn't call me back.

July 29, 2004

I leave more messages on Guy's answerphone. Can he just let me know if he has told them about me, and if so, what was their reaction?

I don't hear from Guy.

I presume the news went down badly.

ACKNOWLEDGMENTS AND BIBLIOGRAPHY

I would like to thank everyone who allowed me to interview them for this story, especially Jim Channon, General Stubblebine, Guy Savelli, and Eric Olson. I harassed Jim so frequently over the past two years—for facts, dates, copies of his manual, reminiscences, verifications of names and places, and so on—that at one point he e-mailed me to say, exhaustedly: "Why do I feel like the copy boy on my own show?" But he always provided the information I asked for.

Jim allowed me to reproduce a drawing from the *First Earth Battalion Operations Manual,* and I thank him for that too.

Although nothing much has been written before about the First Earth Battalion, *Mind Wars,* by Ron McRae (St. Martin's Press, 1984), has a few useful pages on Jim, a paragraph or two of which I have appropriated.

Thanks to Tony Frewin (of *Lobster* magazine and the Kubrick estate) for giving me his copy of *Remote Viewers: The Secret History of America's Psychic Spies,* by Jim Schnabel (Dell, 1997). This book gave me invaluable background information for chapters 5 and 6, as did Francis Wheen's

ACKNOWLEDGMENTS AND BIBLIOGRAPHY

How Mumbo-Jumbo Conquered the World (Fourth Estate, 2004), and my conversations with Skip Atwater and Joe McMoneagle, the two leading players in the Fort Meade psychic spying unit.

Thanks to John Le Carré, who told me to read *In the Time of Tyrants: Panama: 1968–1990*, by Richard M. Koster and Guillermo Sánchez (W. W. Norton & Company, 1991). All anyone needs to know about Panama and military intelligence is contained within that book.

Prudence Calabrese's moving and funny memoir *Intentions: The Intergalactic Bathroom Enlightenment Guide* (Imprint, 2002) helped me to retell the story of her roller-coaster adult life, and I recommend it. Well, the memoir parts I recommend unreservedly, the alien-in-the-bathroom parts I recommend reservedly.

Thanks to Kathryn Fitzgerald Shramek for allowing me to reproduce her late husband's photograph of the Hale-Bopp comet and the "companion" object.

It was a pleasure to watch the brilliant documentaries *Waco: Rules of Engagement* and *Waco: A New Revelation* again. Thanks to the producer, Mike McNulty, for sending them to me. The excerpts from the FBI negotiation tapes I quote in chapter 12 are lifted from these superb films.

I pieced together the story of Frank and Eric Olson primarily through my many conversations with Eric, but some paragraphs have been taken from his friend Michael Ignatieff's *New York Times* article, "What Did the C.I.A. Do to Eric Olson's Father? (April 1, 2001); "The Sphinx and the Spy: The Clandestine World of John Mulholland," by Michael Edwards (*Genii*, April 2001); and from Eric's own painstakingly researched web site, www.frankolsonproject.org. Ignatieff's piece was particularly helpful.

Eric allowed me to reproduce two of his photographs, and I thank him for that. I have been unable to locate Ed Streeky, the copyright holder of the third photograph, the one that appeared in *People* magazine in 1975 and shows the family back at home after meeting President Ford.

My information about Artichoke came from Martin A. Lee and Bruce Shlain's *Acid Dreams: The Complete Social History of LSD, the CIA, the Sixties and Beyond* (Pan, 1985).

Thanks also, as always, to Fenton Bailey, Rebecca Cotton, Lindy Taylor, Tanya Cohen, and Moira Nobel at World of Wonder, and the extremely patient Peter Dale at Channel 4. I couldn't have asked for kinder supporters within the channel than Peter and Tim Gardam, the now retired director of programs, and his successor, Kevin Lygo.

Ursula Doyle, my editor at Picador in London, and Geoff Kloske, my editor at Simon & Schuster in New York, were typically brilliant, as were Adam Humphries, Andrew Kidd, Camilla Elworthy, Stephanie Sweeney, Sarah Castleton, and Richard Evans at Picador, and Derek Johns at A.P. Watt.

Most of all I'd like to thank Andy Willsmore, David Barker, and especially John Sergeant, to whom this book is dedicated. John's research and guidance can be found in every page of this book.

ABOUT THE AUTHOR

JON RONSON is the award-winning author of *Them: Adventures with Extremists* and a documentary filmmaker. He lives in London.

Also available from Jon Ronson

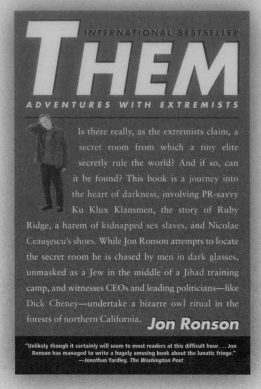

Jon Ronson journeys deep into the lives and minds of extremists—
Islamic fundamentalists, neo-Nazis, Klansmen—and reveals their shared belief
that a shadowy, elitist cabal runs the world. In this internationally bestselling
investigation, by turns serious and hilarious, Ronson explores the fervor and
fanaticism of the search for *Them*.

"Jon Ronson has managed to write a hugely amusing
book about the lunatic fringe."

—Jonathan Yardley, *The Washington Post*

"Often entertaining, more often disturbing . . . [Ronson]
has gotten closer to these people than any journalist I
can think of."

—Ron Rosenbaum, *The New York Times Book Review*